Zdenek Hejzlar, Julie Kilgore, Norman Eke

Technical Aspects of Phase I/II Environmental Site Assessments–3rd Edition

ASTM Stock Number: MNL43-3RD

ASTM International
100 Barr Harbor Drive
PO Box C700
West Conshohocken, PA 19428-2959

Printed in the U.S.A.

Library of Congress Cataloging-in-Publication Data

Hejzlar, Zdenek, 1955-

   Technical aspects of phase I/II environmental site assessments/Zdenek Hejzlar, Julie Kilgore, Norman Eke. — Third edition.
      pages cm
"ASTM Stock Number; MNL43-3RD."
Includes bibliographical references and index.
ISBN 978-0-8031-7043-8 (alk. paper)
1. Hazardous wastes—Risk assessment—Standards—United States. 2. Hazardous waste sites. 3. Groundwater—Analysis. 4. Soils—Testing.
   5. Soils—Classification. 6. Engineering geology—Standards—United States. I. Kilgore, Julie H. II. Eke, Norman. III. Title.
   TD1050.R57H45 2015
   363.7'02—dc23   **2015005740**

Copyright © 2015 ASTM International, West Conshohocken, PA. All rights reserved. This material may not be reproduced or copied, in whole or in part, in any printed, mechanical, electronic, film, or other distribution and storage media, without the written consent of the publisher.

Photocopy Rights

Authorization to photocopy items for internal, personal, or educational classroom use, or the internal, personal, or educational classroom use of specific clients, is granted by ASTM International provided that the appropriate fee is paid to the Copyright Clearance Center, 222 Rosewood Drive, Danvers, MA 01923, Tel: (978) 646-2600; http://www.copyright.com/

ASTM International is not responsible, as a body, for the statements and opinions advanced in the publication.
ASTM International does not endorse any products represented in this publication.

Printed in Mayfield, PA
May, 2015

Foreword

THIS PUBLICATION, *Technical Aspects of Phase I/II Environmental Site Assessments* was sponsored by ASTM Committee E50 on Environmental Assessment, Risk Management and Corrective Action. It is the third edition of MNL43 of ASTM International's manual series.

Acknowledgments

This manual changed from a dream to a reality because of ASTM and the many wonderful professionals associated with this technical organization.

It all began in 1993 when, at the invitation of ASTM's Scott Murphy, my colleagues on ASTM's Technical and Professional Training team and I developed the first Phase I Environmental Site Assessment Training Course. Since then, by teaching the class throughout the United States and throughout the world, we all enjoyed and continue to enjoy the opportunity to share a wealth of environmental knowledge, practical experiences, and helpful business information with each other and our students.

The experience was broadened further when we developed the Phase II Environmental Site Assessment Course, which was completed in February 1999 and has been offered by ASTM since then on a regular basis. Because of the natural progression and close relationship of the two standards, several of the instructors now teach both courses. It was during the development of the Phase II Course that I first conceived the idea of a manual based on the principles of the two standards, but that would also address many of the issues brought up during the classroom sessions. With the encouragement and support of my colleagues, I set out to fulfill that dream.

The first edition of this manual was published in 1999. In 2002, I completed the work with Mr. Shunichi Kamewada on a translation of the manual into Japanese, and I subsequently visited Japan on several occasions to train Japanese instructors in several ASTM Standards. In 2007, the manual was revised. Mr. John Worlund co-authored the edition with Dr. Hejzlar.

In recent years, with the changes in the U.S. Environmental Protection Agency Brownfields programs and related legislative developments, several changes evolved in the industry that drove the need to revise the ASTM Standards and the necessity to edit and update this manual. Ms. Julie Kilgore, who led the ASTM task group that revised the Phase I standard, and Mr. Norman Eke, who led the subcommittee of the task groups that revised the Phase I and Phase II standards, agreed to co-author the third edition with Dr. Hejzlar.

Many of the examples in the book are based on actual experiences of the instructors and students. Because of client confidentiality issues, many of the examples and case studies have been purged of names and references. In this third edition, several of the examples and photographs have been updated and revised. Sincere thanks go to the instructors who provided the materials, encouragement, and expertise during the writing and technical review. Without their help, this manual would not have been possible.

Special thanks also go to ASTM technical professional trainers for drafting the chapter on safety and health and to David Vieau for providing materials presented in the chapter on work plans.

Subject Material Contributors
Jane Bral
Pete Glubiak
John Herbert
John Manzo
James Mauch
Patrick Rhoads
Jay VanHouten
David Vieau
Derrek Verlaan
Jaroslav Solc
James Dismukes
Nicolas Albergo
William R. Weissman
Paul Zovic

—Zdenek Hejzlar

Contents

Foreword	iii
Acknowledgments	v
Introduction	1
Part 1 Phase I Environmental Site Assessment Process	5
1 Scope of Phase I Environmental Site Assessments	7
2 Speaking the Language: Terminology	13
3 Significance and Use of E1527	19
4 Responsibilities	25
5 Records Review	31
6 Site Reconnaissance	37
7 Safety and Health	51
8 Interviews	57
9 Report Preparation	63
10 Dealing with the Nonscope Issues	71
Part 2 Phase II Environmental Site Assessment Process	75
11 Scope of Phase II Environmental Site Assessment	77
12 Expanding the Environmental Vocabulary	83
13 Application and Uses of E1903	89
14 Contracts	95
15 Developing the Scope of Work: Work Plans	101
16 Assessment Activities	107
17 Validation Process and Interpretation of the Results	119
18 Phase II ESA Report	125
19 Dealing with Nonscope Issues in Phase II Assessments	131
Appendix: Forms	133
Building Permit Review	134
Property Reconnaissance	135
ASTM 1527-13 APPENDIX X3—User Questionnaire	139

User Responsibility Statement	141
Owner Interview and Helpful Documents	142
Table of Contents Template Discussion	144

Index 145

Introduction

What Is the Manual About?

This manual is about applying ASTM Environmental Assessment Standards to environmental issues so that the various participants involved in commercial real estate transactions can make sound business decisions.

What This Manual Is Not About

This manual will not teach anyone how to be an environmental consultant, engineer, geologist, or a scientist because it cannot cover or replace years of education and experience or the latest in instrumentation or remediation methods.

Who Would Benefit from Reading It?

ASTM Standards are tools to promote understanding among buyers, engineers, consultants, regulators, owners, attorneys, lenders, Brownfields grantees, real estate brokers, appraisers, economic development agencies, and those involved with environmental insurance.

Commercial transactions involve numerous varied and often competing interests. Even in a simple scenario involving only a seller, buyer, banker, and an environmental professional, numerous conflicting interests abound. The seller and the buyer want a fair price. A fair price is not necessarily the same figure to both parties. All parties perceive any additional cost associated with the transaction as detrimental unless the other party pays for it. If some contamination is discovered on the property, then the buyer may want to use it as a bargaining chip. On the other hand, a buyer may think that environmental professionals make their living by cleaning up properties, using this as an opportunity to generate more business. The lender wants to make the loan, but wants to minimize the risk to the bank if the loan goes bad. One way to minimize risk is to hire a reputable environmental professional to thoroughly check the property. This will cost money, and if the cost is too high, then the buyer might obtain the loan from a competing lender who has more risk tolerance, uses a cheaper environmental professional, or asks no environmental questions at all. The risk of the buyer, who could be liable for the cost of cleaning a contaminated property, is significantly different than that of the banker, who only risks some percentage of the remaining loan balance. Lastly, everyone wants it done yesterday. This manual examines how the ASTM Standards can be used as tools to resolve some of these issues.

How Is the Manual Organized?

Part 1 of the manual is dedicated to the understanding and appreciation of the Phase I Environmental Site Assessment **E1527** Standard [1], including the changes incorporated into the 2013 version. Discussion has been added to explain the relationship between the U.S. Environmental Protection Agency (EPA)'s All Appropriate Inquiries (AAI) regulation [2] and ASTM **E1527**. Part 2 builds on the experience and knowledge gained in Part 1 and covers **E1903** [3], elaborating on the technical business application of the Phase II Standard. The latest version of the standards can be obtained from ASTM.

Reasons for ASTM Environmental Site Assessment Standards

Federal law creates liability for pollution or contamination with hazardous substances which is strict, joint and several. "Strict liability" means that fault is not a prerequisite. It does not matter if the owner was obeying the law when they disposed of the hazardous waste. "Joint" and "several" liability means that any of the parties in the lawsuit may be liable for the entire cost of the cleanup. This provision is referred to as "deep pocket" because those parties that have the money to pay for the cleanup will generally be targeted. The liability is also retroactive, which means that it does not matter when the pollutant was released.

Potentially responsible parties (PRPs) can clean up the sites themselves with EPA or state oversight. For example, if owners find out that the site is contaminated, they can clean it up with either state or EPA's supervision. If they determine that someone else caused the contamination, such as prior owners, then they can sue to recover the cost of cleanup. On many sites, the cleanup costs are (or are perceived to be) so high that the owners may have abandoned the sites. In those instances, EPA and/or the state government can start actions to clean up the sites and then sue PRPs to recover the cost of the cleanup. There is a difference between private parties and the government trying to recover response costs (costs of cleanup). Broadly speaking, a private party must prove that its costs were necessary and that the cleanup action was consistent with the National Contingency Plan (NCP). If the government cleans up the site and sues, then it is presumed that the cleanup action was necessary and was consistent with the NCP.

There were initially three defenses to liability for EPA response costs:

1. Act of God
2. Act of war
3. Third party and the purchaser had no reason to know.

For example, the owner would not be held liable if the contamination of the property was caused by an earthquake (although a court might consider whether the facility was located in an area riddled with known fault lines and preventative measures should have been taken) or if a release was caused by a terrorist strike (although a court might consider whether war had actually been declared at the time the release occurred).

Considering the low probability of these events, the only real defense against liability for cost of cleanup is what was referred to in the ASTM Standards as the Innocent Landowner Defense (ILD). To qualify for the ILD, the purchaser must

1. Acquire the site after the disposal or placement of hazardous substances on the property, and
2. Demonstrate that the purchaser did not know, or had no reason to know, that any hazardous substances were present on the property.

To establish that the purchaser had no reason to know about the contamination, the purchaser must

1. Undertake all appropriate inquiries (AAI) into the previous ownership and uses of the property, and
2. The appropriate inquiries must be consistent with good commercial and customary practice.

The language was left to broad interpretation, and the ASTM Standards were created with the purpose of defining all appropriate inquiries and good commercial customary practice.

The Brownfields Amendments of 2002 provided two additional defenses. These are the Bona Fide Prospective Purchaser (BFPP) and the Contiguous Property Owner (CPO). These two defenses along with the previous ILD are collectively referred to as the Landowner Liability Protections (LLPs).

Petroleum and crude oil have been explicitly excluded from the definition of hazardous substances. However, they are covered by other regulations and are of concern in many commercial real estate transactions; therefore, they are included within the scope of the ASTM Phase I and Phase II Standards.

The Phase I and Phase II Standards are continuously evolving documents. As the application of the standards changes, case law develops, and the industry evolves, the members on the ASTM Committee review the developments and work to incorporate applicable changes into the standards through the revision process. Environmental professionals and those who contract with environmental professionals (the "users") are strongly encouraged to stay current on the latest revisions and modify their procedures to reflect applicable changes.

Regulatory Background

Congress, citing their constitutional power to regulate commerce amongst the states, has enacted broad regulatory statutes delegating implementation to administrative agencies. Adopting detailed regulations takes many years, and the scientific data intended to justify the regulations may be uncertain and subject to conflicting interpretations. Once regulations are adopted, parties that are dissatisfied with the result may file legal challenges to them. Agencies frequently publish guidance documents or issue interpretative letters. Agency regional offices sometimes issue policy statements that may govern the conduct of businesses within their region, especially on subjects that the agencies' headquarters may not have addressed. There are also state laws and regulations, and the federal government often defers to state programs so long as they believe that the state program is at least as stringent as the federal program. Environmental law is a complex and dynamic subject that continues to be very much in flux. Throughout the manual, we will present several examples of case law outcomes.

The main federal regulations are

Year	Regulation
1970	U.S. EPA created.
1970	Clean Air Act (CAA) gave EPA the power to establish clean air standards. Amendments followed in 1974, 1977, and 1990. In addition to establishing ambient air pollution standards such as those regarding vehicle emissions, EPA also establishes standards for point sources, such as factories and power plants.
1972	Federal Water Pollution Act (FW-PCA), predecessor to the Clean Water Act.
1976	Resource Conservation and Recovery Act (RCRA) and the Toxic Substance Control Act (TSCA). RCRA is designed to manage industrial wastes from the point of generation to their final disposal location. It is often referred to as the "cradle to grave" management process for hazardous wastes. The act requires the generators and transporters to identify, quantify, and characterize their hazardous wastes. It also specifies performance standards and permit systems for treatment storage and disposal (TSD) facilities. Most important for those involved in real estate transactions, all petroleum products are regulated under the RCRA.
1976	The TSCA regulates the manufacture, use, and disposal of chemical substances and requires the substances to be tested to determine whether they pose an unreasonable risk to human health or the environment. For example, the TSCA introduced extensive regulation covering polychlorinated biphenyls (PCBs).
1977	Clean Water Act (CWA), under which the EPA controls pollutants entering all surface waters, including lakes, rivers, estuaries, oceans, and wetlands as they define them.
1980	Comprehensive Environmental Response, Compensation, and Liability Act (CERCLA), commonly known as the "Superfund Law," created the need for environmental site assessments for commercial property transactions. It generated a "blueprint" on how the hazardous substances are to be cleaned up, known as the NCP. If lenders could be perceived by the court as participating in management, then the lenders were held liable for clean up of the properties. CERCLA created a fund for clean up of the dangerous sites that is based on evaluation and priority listing, which was initially from taxes on the chemical industry and 42 commercial chemicals.

1984	RCRA amended by the Hazardous Solid Waste Amendments (HSWA), which aimed at the protection of groundwater by new requirements for managing and treating small quantities of hazardous wastes such as those generated by auto repair shops and dry cleaners and stricter requirements for municipal solid waste and industrial landfills.
1985	Asbestos Hazard Emergency Response Act (AHERA) requires schools to identify and respond to their asbestos problems. The Asbestos School Hazard Abatement Act (ASHAA), passed in 1984, provided interest-free loans to schools for asbestos control projects.
1986	Superfund Amendments and Reauthorization Act (SARA). The CERCLA/SARA liability provisions are very broad and require EPA to recover the cost of any cleanups from the "responsible parties," which include all past and present owners and operators of the property. The details of this liability recovery and the defenses to it led to the development of the ASTM Phase I Environmental Site Assessment Standard, described in more detail in the Reasons section for ASTM Phase I.
1986	Emergency Planning and Community Right-to-Know Act (EPCRA) requires hazardous chemical emergency planning by federal, state, and local governments; Indian tribes; and industry. It requires industry to report on the storage, use, and releases of hazardous chemicals to federal, state, and local governments.
1990	Hazardous Materials Transportation Uniform Safety Act (HMTUSA) regulates the transportation of hazardous materials in intrastate, interstate, and foreign commerce.
1990	Pollution Prevention Act (PPA) authorized EPA to encourage the adoption of source reduction techniques by businesses using matching grants and information exchange on the topic.
1992	EPA created a rule providing a Secured Creditor Exemption from CERCLA liability.
1994	The 1992 EPA Secured Creditor Exemption was declared invalid.
1996	CERCLA was amended to provide a Secured Creditor Exemption creating a "safe harbor" for many customary lending practices, although a lender who exercises too much control may lose this protection. Lenders could still lose all or part of the remaining loan balance if the borrower is held liable for cleanup; therefore, they often conduct environmental due diligence on the basis of their risk tolerance. The risk tolerance of the borrower may be significantly different than that of the lender.
2002	Small Business Liability Relief and Brownfields Revitalization Act (Pub. L. 107–118, 115 Stat. 2356, "the Brownfields Amendment") [4] required EPA to develop regulations establishing standards and practices for how to conduct "all appropriate inquiries," and Congress included a list of criteria that EPA must address in the regulations establishing standards and practices for conducting all appropriate inquiries.
2005	EPA issued their All Appropriate Inquiries Final Rule (AAI).
2013	AAI Final Rule Amended to reference revised ASTM Standard.

In addition to the federal regulations discussed here, many states and local governing bodies within each state have adopted environmental regulations that mirror the federal statutes. Local regulations can often be stricter than federal laws or deal with issues not covered by federal regulations. Throughout this manual, we will stress the importance of local knowledge with respect to environmental assessments and the need for the environmental professionals to become acquainted with state and local requirements.

References

[1] ASTM Standard **E1527**, Standard Practice for Environmental Site Assessments: Phase I Environmental Site Assessment Process, *Annual Book of ASTM Standards,* ASTM International, West Conshohocken, PA, 2013.

[2] U.S. Environmental Protection Agency, "All Appropriate Inquiries," Final Rule 40 CFR Part 312.

[3] ASTM Standard **E1903**, Standard Practice for Environmental Site Assessments: Phase II Environmental Site Assessment Process, *Annual Book of ASTM Standards,* ASTM International, West Conshohocken, PA, 2011.

[4] Small Business Liability Relief and Brownfields Revitalization Act (Brownfields Bill), 42 USC §§9601 *et seq.*, 2012.

Part 1

Phase I Environmental Site Assessment Process

Chapter 1 | Scope of Phase I Environmental Site Assessments

1.1 History of the ASTM Standard

The first ASTM E1527 Standard Practice was published in 1993, although the actual development of that standard began in the late 1980s. As will be explained in more detail, "Superfund" liability was the driving force behind the development of the standard. The actual law, titled the Comprehensive Environmental Response, Compensation, and Liability Act (commonly referred to as "CERCLA" or "Superfund") was passed in December 1980 [1]. There was a significant amendment in October 1986 titled the Superfund Amendments and Reauthorization Act, commonly referred to as "SARA."

It took the commercial real estate industry approximately 10 years to develop an understanding of the requirements of CERCLA, and many did not appreciate the broad liability impacts or due diligence obligations until the ASTM Phase I Environmental Site Assessment (ESA) process was published. Thus, the learning curve was quite steep for those with limited involvement in the Phase I process. In today's commercial real estate market that is no longer the case. Buyers, developers, lenders, and brokers are now more likely to be familiar with a "Phase I," even if they may not be fully aware of the specific details of the process. Nonetheless, there are exceptions. With more foreign corporate investments there may be users that do not appreciate the broad liability impact or due diligence obligations and may rely on consultants for specific advice and guidance.

This book is designed to answer the common questions that arise in the application of the ASTM E1527 Standard Practice. However, before we start talking about the application of the E1527 Standard, a little history would be helpful.

The Phase I ESA standard was developed to provide a framework for conducting "all appropriate inquiries," a phrase originating from the 1986 SARA previously mentioned. As discussed earlier, the first E1527 Standard was published in 1993. The overarching objective of the standard was to lay out the standardized components that must be included in a Phase I ESA, but leaving the selection of specific resources to the judgment of the environmental professional conducting the assessment.

> **E1527-13:** "Objectives guiding the development of this practice ... are (1) to synthesize and put in writing good commercial and customary practice for environmental site assessments for commercial real estate, (2) to facilitate high quality, standardized environmental site assessments, (3) to ensure that the standard of appropriate inquiry is practical and reasonable, and (4) to clarify an industry standard for appropriate inquiry in an effort to guide legal interpretation of CERCLA's innocent landowner defense."

A revised version was published shortly thereafter in 1994. The 1994 revisions were of an editorial nature and concentrated on the Historical Review section. In fact, the revisions would probably go unnoticed by a casual reader of the standard. The next revision was in 1997, and it also was largely editorial, with the most significant modifications made in the Data Search sections. These changes were agreed upon by the responsible ASTM task group to clarify the original intent of the document and were based on industry input and experience from using the document since 1993.

The next revision was in 2000, which was the most significant revision up to that point. It is in the 2000 revision that we see the introduction of the definition of a "material threat," the concept of a historic recognized environmental condition (HREC), and discussions related to "activity and use limitations" (all of which are discussed in more detail in subsequent chapters). There was also nearly a total rewrite of the Report section to make it clear that the Phase I ESA was more than a mere "data dump" of information and that the environmental professional was required to provide opinions regarding all of the findings.

Up through the early 2000s, the Phase I document was used primarily in the commercial real estate practice. There was very little involvement on the part of U.S. Environmental Protection Agency (EPA) or state environmental agencies. Some states had initiated requirements or registries of people who performed Phase I services. Some regulators had become aware of the Phase I process and had seen or even requested Phase I documents related to a specific release site or suspected release site. However, by and large, the Phase I ESA process typically did not rise to the level of regulatory involvement.

All of this changed on January 11, 2002, when President George W. Bush signed the Small Business Liability Relief and Brownfields Revitalization Act, more commonly called the Brownfields Amendments.

One of the directives of the Brownfields Amendments required EPA to develop regulations for conducting all appropriate inquiries. The Brownfields Amendments cited the 1997 version of the ASTM **E1527** Standard and stated that it should be used until EPA completed the process of developing a regulation for conducting all appropriate inquiries. This put the EPA in a position of having to develop a rule to potentially replace a consensus standard practice that had been commonly applied within the real estate industry for over a decade.

The Brownfields Amendments also introduced two new legal defenses. In addition to the Innocent Landowners Defense (ILD), which had previously been available as a defense to CERCLA liability, it added the Bona Fide Prospective Purchaser (BFPP) and the Contiguous Property Owner (CPO) Defenses. These three defenses are collectively referred to as the "Landowner Liability Protections" (LLPs) to CERCLA.

The BFPP LLP allows a buyer to purchase a property with knowledge that the site is contaminated. This is important because before 2002, the Phase I process was designed with the understanding that only the ILD was available and would only be available if, at the time of purchase, the buyer did not know and had no reason to know that the property was contaminated. One of the critical components of the Brownfields Amendments that changed the face of the Phase I ESA industry was revisions that outlined the approach to owning, developing, and managing contaminated or potentially contaminated properties. Under the provisions of the 2002 Brownfields Amendments, along with this new ability to buy contaminated property with knowledge, there also came "continuing obligations" that take effect upon acquisition of the property. These continuing obligations will be discussed again in more detail later in this document. The key point to make is that although these continuing obligations are critical to maintaining the defense, neither AAI nor the ASTM Phase I Standard fully addresses their implementation.

Because it became clear during the EPA rule-making process that the implementation of the EPA All Appropriate Inquiries Final Rule (AAI) was going to address many of the same issues that confronted the drafters of the various versions of the ASTM **E1527** Standards, the ASTM document became a useful tool. Although the EPA was going through the process of developing AAI, the ASTM task group was also reconvened to make the changes necessary to comply with the anticipated AAI regulation. In 2005, EPA published the AAI regulation 40 CFR Part 312 as a relatively nonprescriptive performance-based process, referencing the 2005 version of the **E1527** as an optional tool that could be used in the performance of AAI. (In 2009, the EPA also referenced the 2008 version of the **E2247** Standard Practice for Phase I Environmental Site Assessments for Forestland or Rural Property as being compliant with AAI.)

Because all ASTM standards have a maximum shelf life of 8 years, in 2010 the ASTM **E1527** task group began a careful review of the 2005 standard and gathered industry feedback regarding the application of **E1527** for meeting the technical and practical needs of AAI and the commercial real estate community. Two primary principles guided the revisions to ASTM **E1527**: (1) clarify language in which there was inconsistency in interpretation and (2) strengthen the deliverable that is produced at the end of the Phase I ESA process.

The task group looked at terms that were being interpreted quite differently, often resulting in very different end results. The task group also revisited terms that were being applied in a different context than was originally intended. In many cases, the variability in application was not necessarily incorrect, but it led to confusion and inconsistency in how information was presented to the end user.

As the task group moved through the ASTM revision process, EPA also participated so as to ensure that the proposed revisions would not conflict with the requirements of the CERCLA statute or the AAI regulation. Once the **E1527** task group successfully worked through ASTM's consensus process for approval of revisions, a revised standard, **E1527-13**, was published in November 2013, and in December 2013 EPA modified the AAI regulation 40 CFR Part 312 to reference the updated **E1527** standard.

1.2 Introducing the Players

Phase I ESAs involve "users" and "producers" as the two basic players in an **E1527** Standard Practice for Phase I Environmental Site Assessment Process [2]. The producer would be the environmental professional, the qualified professional who conducts the tasks normally associated with completing a Phase I or who remains in responsible charge of those who perform the task elements. Users include anyone who may be using the standard. This could include lending organizations, developers, property owners, lessors, grantees, or facility operators.

Another portion of the relationship between the user and the environmental professional is defined by the needs and professional responsibilities of the players. There are numerous ways for the relationship to develop. By examining one possible scenario, we will introduce the typical players in a Phase I ESA. We will use the scenario to point out numerous advantages that **E1527** affords the players.

1.2.1 THE LENDER

It is February and a busy time of year for a commercial real estate lending officer in a fast-growing Florida town. The lender is ready to close on a $2 million loan for a large industrial warehouse by the end of the month. A corporate directive requires the performance of a Phase I ESA on all property transactions with the loan value in excess of $1 million. The document references concerns regarding the lending of money on contaminated properties. After a three-bid policy (yet another procedure at the bank), the lender searches the Internet and calls three environmental service firms and requests Phase I proposals.

The following morning, the lender has three proposals on his desk. The first includes two pages describing the services, five pages of disclaimers, and a preliminary cost estimate of $10,000. The second proposal includes a five-page description of the scope of services, one page of fine print disclaimers, and a cost estimate of $3,000. The third proposal simply reads "Phase I Audit on property described above $500."

The lender reviews the information and begins to consider his options. He is well aware of the fact that his client has a choice in selecting a bank. The bank will pass through the cost of the Phase I ESA to the client as part of the closing costs, but the client will probably want to know how (and why) the money was spent. Because the client is paying for the Phase I ESA, he also has some input in who is hired. The lender would like to disregard the lowest bid because he is suspicious of the very low price; he asked for an "assessment," and the bid is for an "audit" and he suspects that the two are entirely different. A firm he has used in the past submitted the middle bid, and they are proposing to look into several items that were familiar to him from past Phase I ESA reports he had seen. The highest bid includes a laundry list of services, of which many will not have any effect on his assessment of the loan associated with the property. What does he do now? Whom does he trust, and how is he going to justify the cost of the assessment to the buyer (client) applying for the loan?

The ASTM Standard defines this player as the user. When he requested the proposals, he could have specified that he wanted the Phase I ESA performed in accordance with the scope of **E1527**. The lender could then use the scope of the standard to compare it with the proposed scope of work, and he could have directed the engineering firms to include the standard-required elements (thus likely raising the cost of the $500 bid) and eliminate all of the other unrelated services (perhaps reducing the cost of the higher bid). This would have streamlined the bidding process and enabled him to receive comparable bids. Lastly, he could have discussed the standard with the buyer and guided the buyer in deciding which environmental professional to hire.

It is not uncommon for lenders to be the primary drivers for conducting Phase I ESAs, not because lenders are concerned about CERCLA LLPs (at loan origination or refinance, lenders are not property owners; therefore, they are covered by the Secure Creditor Exemption) but because the Phase I ESA process is a useful tool that a lender uses to evaluate the collateral of their borrower, much like an appraisal or a building inspection. Larger lending groups have internal environmental review managers who are well versed in Phase I ESAs and CERCLA LLPs. However, many of the smaller lenders and community banks are not, and they can easily be swayed by low-price services that do not meet the requirements of **E1527** or the AAI regulation. Furthermore, the 1996 Secured Creditor Exemption, which partially limits lender's CERCLA liability, allows the lender further latitude in their approach to Phase I ESAs.

1.2.2 THE BUYER

For the purposes of a CERCLA LLP, a buyer of commercial property is the true end user of a Phase I ESA, but the traditional purchaser of a small piece of commercial real estate is rarely informed about or inclined to spend the money on this thing called a Phase I ESA. A property transaction may involve many million of dollars, but it can be difficult to see the value in Phase I ESA, let alone soil and groundwater sampling in the event that the Phase I ESA identifies a potential concern.

The risk tolerance of the lender and that of the borrower (buyer) are likely to be very different. Thus, the borrowers need to take care when a lender selects a Phase I provider who "understands the lender's risk tolerance" because it is the borrower who becomes the new owner and who could be made to pay the remediation costs whereas the lender simply risks some percentage of the remaining loan balance.

E1527 can be used to explain the concepts of due diligence, potentially responsible parties, all appropriate inquiries, and other relevant concepts covered by the standard. The buyer would then have a better understanding of the implications associated with the transaction and the reasons for spending the money on the ESA. The decision regarding whether to perform a Phase I ESA at all, and the level of inquiry the buyer selects, boils down to his and his lender's risk tolerance. They can choose to accomplish the due diligence that satisfies AAI or be comfortable with the risk of potentially losing their investment and owning the liability for cleanup.

1.2.3 THE ENVIRONMENTAL PROFESSIONAL

The standard defines this individual responsible for the Phase I ESA as the environmental professional (EP). It is often the EP who uses the **E1527** to educate the buyer and the lender about the significance of the Phase I ESA activities and to justify the costs for this service. The EP responsible for the Phase I ESA must meet certain education and experience requirements that will be discussed in later chapters.

The parties may wish to comply with AAI because they believe it is good practice, they want to make sure they qualify for LLPs, or they may be involved in a project related to Brownfields grants (as discussed below). In any of those instances, one can comply with AAI without using the ASTM Standard. The principal difficulty is determining what extent of research and what level of detail is required. There are also several areas in which ASTM is more prescriptive than AAI. Because the ASTM Standard has been generally considered representative of good commercial and customary practice, there may be some liability in using AAI criteria that fall short of ASTM criteria. The ASTM Standard is often used as the benchmark by the user and the courts.

It is important to recognize that the ASTM Phase I process is conservative in that it will identify many issues that would never rise to the level of Superfund liability. Many users do not really care about that Superfund liability; rather, they just want to identify issues that might cost them money, affect their development plans, or result in some kind of non-Superfund liability, for example, a third-party neighbor. The ASTM Phase I Standard discusses non-scope business environmental risk that typically would not be related to Superfund liability. Considerations beyond Superfund liability make it even more important to ensure that the user and the EP are very clear about the scope of work and the process being used to satisfy that scope of work.

1.2.4 THE REGULATOR

The Brownfields Amendments also direct the EPA and the grantees involved in the EPA Brownfields grants program to use AAI (and by inference ASTM **E1527**) in administering the Brownfields grant program. Most state environmental agencies must document that they are complying with AAI in administering the grants. As a practical matter, the EPA, the states, and grantees, similar to the commercial real estate industry, are specifying the ASTM **E1527**

Standard to define their contract scope of work. The Brownfields grant programs drastically expanded the applicability of Phase I ESAs to circumstances in which regulators were already involved at a site, and regulators are now often closely involved with, and at times can be an integral part of, the Phase I ESA process.

1.3 Scope of E1527

Before the publication of the first edition of the ASTM Standard on Phase I ESAs, the users and the EPs had to develop the ESA scope on their own. Consequently, Phase I assessments came in many variations. On one hand, the users were seeing Phase I reports without any interviews or historical research. Some reports consisted of "drive-by" photo documentation without any evidence of the EP ever setting a foot on the property. On the other hand, a Phase I assessment sometimes included issues such as radon, electromagnetic radiation, compliance issues, fire hazards, and structural evaluations of buildings, which are not CERCLA issues. In summary, the term "Phase I" or "Level 1" or "Pre-Acquisition Environmental Audit" meant different things to different people. The E1527 Standard was developed by ASTM through its Committee E50 on Environmental Assessment to establish consensus-based (agreed upon) and uniform (standardized) site assessment practices. The intent was to define some of the due diligence responsibilities of the participants (our referenced players) in a commercial real estate transaction. Through standardization, the E50 Committee made it possible for the Phase I ESA to mean the same thing to different people.

Let us examine the Scope section of the standard and see how the EPs can apply it to their work. This section of E1527 defines three essential issues: (1) the purpose of the standard, (2) the objectives of the standard, and (3) considerations beyond the scope of the standard. Let us separately look at each issue.

1.3.1 PURPOSE

In the language of the standard, the purpose of E1527 is to define a practice for conducting Phase I ESAs. The EP is going to be considering only a range of contaminants covered by CERCLA as well as petroleum products (this is one area in which the E1527 is more stringent than AAI because of a "petroleum exclusion" to CERCLA liability). The Phase I standard is designed to permit the user to satisfy one of the requirements for the LLPs to CERCLA liability; that is, the practice that constitutes "all appropriate inquiries into the previous ownership and uses of the property consistent with good commercial and customary practice" as defined at 42 U.S.C. § 9601(35)(B). The goal of the Phase I ESA is for the EP to identify recognized environmental conditions (RECs).

1.3.2 OBJECTIVES

The objectives spelled out in Section 1.2 of the standard were the guiding objectives during the development of the standard (see **Fig. 1.1**). They included having a written document with a standardized level of quality assurance requirements that were practical and reasonable. Another objective for developing the standard was to clarify what "appropriate inquiry" meant so that

FIG. 1.1 Objectives from ASTM **E1527-13**.

> Objectives—Objectives guiding the development of this practice are (1) to synthesize and put in writing good commercial and customary practice for environmental site assessments for commercial real estate; (2) to facilitate high-quality, standardized environmental site assessments; (3) to provide a practical and reasonable standard practice for conducting all appropriate inquiries; and (4) to clarify an industry standard for all appropriate inquiries in an effort to guide legal interpretation of the LLPs.

it could be used as a guidance document in the legal interpretation of CERCLA LLP issues.

1.3.3 CONSIDERATIONS BEYOND THE SCOPE

E1527 has limitations. Some of the limitations are listed in the Purpose section of the standard. Other limitations appear elsewhere in the standard and will be discussed in the following chapters. The EP is sometimes asked, or is inclined, to include issues in the Phase I ESA that are beyond the scope of E1527. The user may not be interested in LLPs and is using the Phase I ESA for business decisions. Whether or not nonscope issues should be included as part of the Phase I ESA work will depend on how the EP defines the scope of services with the user.

1.3.4 ESTABLISHING SCOPE OF WORK WITH YOUR CLIENT

During the contracting stage in anticipation of the performance of a Phase I ESA, the EP and the user have three basic options:

1. Use **E1527** to define the scope.
2. Define a scope different from **E1527**.
3. Use the **E1527** scope and define any additional services beyond the scope.

The first option, using E1527 to define the scope of the assessment, simplifies the scoping process. The scope agreement may simply state, "We agree to perform Phase I ESA in accordance with the scope and limitations of the ASTM E1527-13." This will work if the client is familiar with the standard and has sufficient understanding of the implications associated with such a statement. However, it is often helpful to provide the client with more information. This can be a relatively simple task to accomplish by using the language directly out of the standard. **Figure 1.2** is an example of an expanded statement of the Phase I ESA scope, providing the user with significantly more information.

The scope statement in **Fig. 1.2** uses terminology defined by the standard. The task group did not create any new or unique terminology that would require lengthy review by a group of attorneys. The process of standardization has already accomplished that. The task group members involved in the standard development are experts in their respective fields, including scientists, engineers, and attorneys. Over several years, they refined the document until all could agree to publish it. Having said that, it is recognized that use of the standard is voluntary. This means that

FIG. 1.2 Example of Phase I ESA statement of scope.

> We would conduct this assessment in general accordance with the scope and limitations of the *Standard Practice for Environmental Site Assessments: Phase I Environmental Site Assessment Process* as presented in ASTM International's (American Society for Testing and Materials) **E1527-13**. The purpose of the Phase I ESA is to identify, to the extent feasible pursuant to the processes prescribed in ASTM **E1527-13**, recognized environmental conditions in connection with the property. In accordance with ASTM **E1527-13**, the Phase I ESA would consist of four components: (1) Records Review, (2) Site Reconnaissance, (3) Interviews, and (4) Report.
>
> There may be environmental issues or conditions at a property that parties may wish to assess in connection with commercial real estate that are outside of the scope of the ASTM **E1527-13** practice. Several nonscope considerations include (but are not limited to) asbestos-containing building materials, lead-based paint, wetlands, or radon. No implication is intended as to the relative importance of inquiry into such nonscope considerations, nor is this list intended to be all-inclusive. No nonscope considerations have been requested, and none are included in this assessment.

the "players" do not have to use it to conduct Phase I ESAs. The user and the EP may believe that they can develop a more appropriate scope.

The second option, defining a scope different from **E1527**, which could include using the EPA AAI with no reference to ASTM, is a task that should not be taken lightly. It poses many pitfalls for the parties involved and generally creates weaker documents than those written using the standard. The publication and wide acceptance of the ASTM Standard made this procedure even more difficult because the parties in the transaction eventually may have to defend the scope they chose against the prescribed scope of the standard. The following chapters bring some of these difficulties and pitfalls to light.

The third option, using the **E1527** scope and defining additional services, is a common method used to meet the user's needs. Some users and EPs refer to the expanded scope of services as "Expanded Phase I" or a "Phase I Plus." How the additional services are incorporated into the scope of work is a business decision that is governed by unique aspects of the particular transaction. This book presents some of the general concepts that may be used.

Including out-of-scope items in the scope of **E1527** is common, but it can complicate the process. To minimize confusion, the nonscope issues can be addressed in a separate addendum or under a separate contractual agreement. At a minimum, any additional services should be clearly outlined and reflect a mutual agreement to ensure that the user is provided with the desired services and the EP is not exposed to liability for an unlimited range of issues that might affect the property.

1.3.5 UNDERSTANDING LIMITATIONS

E1527 is often referred to as a "minimum standard," which does not mean that the standard is in any way inadequate. In his discussion on the issue of minimum standards, attorney Steven Hansen notes that all standards, by definition, are minimum standards, and because standards are the consensus of the best minds on the subject, it is impossible to create anything but a minimum standard [3]. However, the needs of the client may dictate a significantly higher or specialized level of inquiry, and the scope of the assessment may have to be expanded to meet those needs. For example, the reasons for expanding the scope may be the user's need to answer questions on issues of radon, wetlands, lead, compliance, reporting issues, and asbestos, just to name a few.

E1527 was designed to address the LLPs to CERCLA. However, this standard is often used for making other business decisions. Consequently, it is often used for a purpose for which it was not designed. This creates unique limitations and new liabilities that tend to be very specific to the transaction. Users and EPs need to proceed with caution when **E1527** is used for business decisions.

E1527 does not provide any guidance beyond the defined scope. **E1527** also does not address any of the safety issues that may be associated with ESAs. **Chapter 7** on safety and health discusses some of these issues.

ASTM has been approved as a process for demonstrating all appropriate inquiries, but AAI and ASTM constitute only part of the requirements for establishing one of the LLPs to CERCLA. Those seeking to perfect their position should take independent actions beyond a Phase I assessment, including knowledge and understanding of continuing obligations. Additional requirements have resulted from the passage of the Brownfields Amendments. Before that time, there was a requirement for documentation that addressed sale price when compared with market value as well as identifying specialized knowledge with respect to a property transfer. However, as a practical matter, the users relied on the Phase I document to satisfy the additional requirements for CERCLA liability protection. The Brownfields Amendments resulted in significant changes. First, they require the user to become more involved in the process, and that involvement is incorporated in the 2005 and 2013 versions of the **E1527** Standard. The Brownfields Amendments also identified continuing obligations, which begin after the acquisition of the property. Those obligations are not addressed in ASTM or AAI. **Figure 1.3** is an excerpt of the specific language included in Section 1.1.3 of **E1527** that alerts the users to this issue.

As we will discuss in later chapters, the decisions associated with the purchase of the property and the level of inquiry selected depend on the risk tolerance of the user. Most of the continuing obligations are likely to be encountered in situations involving a BFPP. This is because those sites have identified contamination on site that may need to be addressed and will need to be properly managed.

1.4 Summary

In this chapter, we provided a brief history of the Phase I standard, discussed the effect of the Brownfields Amendments and AAI, and introduced some of the terminology used in the standard. We limited our discussion to just four types of players in the Phase I ESA. These were the lender, the buyer, the EP, and the regulator.

FIG. 1.3 Excerpt from Section 1.1.3 of **E1527**.

> *CERCLA Requirements Other Than Appropriate Inquiries*—This practice does not address whether requirements in addition to all appropriate inquiries have been met in order to qualify for the LLPs (for example, the duties specified in 42 U.S.C. §9607(b)(3)(a) and (b) and cited in Appendix X1, including the continuing obligation not to impede the integrity and effectiveness of activity and use limitations (AULs), or the duty to take reasonable steps to prevent releases, or the duty to comply with legally required release reporting obligations.

We identified the scope of the standard and how it can be applied to establishing the scope of work with your client. Finally, we also touched on some of the limitations of the standard.

This chapter establishes the foundation for the rest of this book. To those unfamiliar with **E1527**, some of the language used thus far may seem ambiguous, insufficiently defined, or even abstract. As with any language, the effect of **E1527** cannot be truly appreciated until the "players" become fluent in its use. Remember, the goal of standardization is "the same meaning to different people." The manner in which the standard begins to accomplish that goal is through common terminology. In the next chapter, which addresses terminology, the language of the Scope section of **E1527** will begin to take on a clearer meaning that is eloquent in its simplicity.

References

[1] Comprehensive Environmental Response, Compensation, and Liability Act of 1980 ("CERCLA" or "Superfund"), as amended by Superfund Amendments and Reauthorization Act of 1986 ("SARA") and Small Business Liability Relief and Brownfields Revitalization Act of 2002 ("Brownfields Amendments"), 42 U.S.C. §§9601 *et seq.*

[2] ASTM Standard **E1527**, Standard Practice for Environmental Site Assessments: Phase I Environmental Site Assessment Process, *Annual Book of ASTM Standards*, ASTM International, West Conshohocken, PA, 2013.

[3] Hansen, S. W., "Using Standards in Defending Product Liability Cases," *ASTM Standardization News*, Feb. 1996, p. 40.

Chapter 2 | Speaking the Language: Terminology

2.1 Reasons for Standard Terminology

Standardized terminology enables regulators, users, and producers to speak the same language without having to explain the meaning of every technical term used. When the players in a commercial real estate transaction utilize common terminology, it speeds up the communication process, resulting in an efficient use of time and resources.

In the last chapter, we introduced a simple scenario of four Phase I Environmental Site Assessment (ESA) players: the lender, the buyer, the environmental professional (EP), and the regulator. In a real-life commercial transaction, there are often numerous other interested parties, such as lawyers, real estate professionals, appraisers, trustees, grantees, developers, and others. In addition, the term "environmental professional" may include an engineer, a scientist, a biologist, or a geologist. All of these players use technical jargon associated with their specific profession or specialization. For example, when asked to define "construction debris," these individuals readily provide numerous and varying definitions. In some of the Phase I reports, "construction debris" is used synonymously with the term "demolition debris." This becomes a problem when one of these individuals writes a report and different interpretations begin to surface. Standardization defines common definitions for all Phase I players. **E1527** [1] purposely draws a distinction between these two terms in the definitions found in Section 3.2. They read as follows:

- *Construction debris:* Concrete, brick, asphalt, and such other building materials discarded in the construction of a building or other improvement to property.
- *Demolition debris:* Concrete, brick, asphalt, and other such building materials discarded in the demolition of a building or other improvement to property.

Although very similar, the two terms can have significantly different implications with respect to the subject property. The distinction can provide information with respect to age and content of the materials and activities (construction, demolition, or dumping) that occurred at the property. The photograph in **Fig. 2.1** depicts demolition debris that included piping and remnants of pits or sumps from an industrial facility. The age and content of the materials are not known, and it may be difficult for the EP to form opinions with respect to the potential effect of the backfill on the property.

The photograph in **Fig. 2.2** depicts construction debris associated with the recent completion of a new building. This pile of materials does not significantly affect the environmental conditions of the property, and the age of the materials can readily be determined.

Section 3 of **E1527** was developed through the same consensus process described in **Chapter 1**. That means that hundreds of industry professionals (users, producers, attorneys, regulators, and academia) contributed to and agreed on over 100 definitions. Once the standard is published, everyone who performs or uses the Phase I ESA benefits from applying the same terminology. In the event of a future issue, an interested party will locate the Phase I report and begin to scrutinize it in great detail. Words and phrases are examined with respect to that issue and compared back to the standard that was utilized. When terms are used that are slightly altered or misapplied, the benefit of using a consensus-based standard is lost.

Let us look at an example in which things went wrong and resulted in a litigation claim. At issue in this case was the definition of the term "practically reviewable." The attorney's definition varied from the definition in the standard.

After acquiring a piece of property, the purchaser started construction of a warehouse. Near the property boundary, the construction crews encountered petroleum-contaminated soil. The contamination came from an abandoned leaking underground storage tank on the neighboring property. The purchaser sued the EP. Handing the EP a copy of a document from the public agency file, the attorney asks: "Mr. EP, in your report, how could you have missed the agency information that shows the adjoining property had a leaking underground storage tank? This agency's letter is public information, readily available, easily reviewable, and indicates the strong possibility of migration onto subject property." The EP had not seen this letter before today. The EP reviews the agency letter, looks over his report, searches through his notes, pulls out his copy of the standard, and flips through it. After several minutes, wiping cold sweat from his brow, he turns to the attorney and explains, "According to my notes the information kept by this particular agency was only chronologically organized. The only way to

FIG. 2.1   Demolition debris from off site.

Source: Provided by J. Kilgore.

find this document would be to thumb through hundreds of pages of information that were only organized by the date of the record. As defined in Section 3.3.24 of the standard, this information was not practicably reviewable."

Familiarity with **E1527** terminology, appropriate application of it during the Phase I ESA process, and good documentation helped the EP in our example to ensure the validity and strength of his Phase I ESA report.

From a historical perspective, the members of the E50 committee consisted of scientists, engineers, biologists, architects, geologists, real estate developers, consultants, regulators, state officials, lawyers, lenders, investors, property owners, and others.

FIG. 2.2   Construction debris from a new building on site.

Source: Provided by BAI; graphics assistance by GRT.

All of these interested parties did not necessarily agree on all of the issues, but they voted to publish a standard that included standardized commonly defined terms. The terms are an integral part of the standard and are critical to an understanding of the Phase I ESA practice and its use.

2.2 Terminology of the All Appropriate Inquiries Final Rule

Before discussing ASTM **E1527** terminology, let us take a brief look at the U.S. Environmental Protection Agency (EPA) All Appropriate Inquiries Final Rule (AAI) terminology. Only eight terms are defined in Section 312.10 of the EPA AAI.

The following are defined in AAI and were incorporated into the Definitions section of ASTM **E1527**:

- Abandoned property
- Adjoining property
- Data gap
- Good faith
- Institutional controls

The other three terms that are defined in AAI but not specifically defined in the ASTM Terminology section are defined elsewhere in the ASTM document.

Date of acquisition or purchase date is discussed in Section 4 and is footnoted in this same section to cite AAI. The ASTM Standard goes on to explain that if there is no transaction involving an acquisition, then the date is the same as the date of the intended transaction.

The other two terms are also covered in the ASTM Standard. The definition of EP is taken directly from the AAI and is incorporated by reference in the Terminology section and as Appendix X-2 of ASTM **E1527-13**. The AAI-defined term "relevant experience" is used in the definition of an EP and is also included in **E1527-13** Appendix X-2.

Note that many ASTM concepts are used in AAI but with different terminology. Although the ASTM Standard mirrors the eight terms and definitions from AAI, AAI could not cite directly from **E1527**. When the AAI was being drafted in 2003, it was not yet clear whether EPA would reference the ASTM **E1527** Standard in its final rule. Therefore, when EPA wanted to use an ASTM term, copyright issues required that the EPA would have had to cite the ASTM Standard as a source. Because the EPA was not yet prepared to cite ASTM as a source in the EPA rule, different language had to be developed. As a consequence, there are many terms in AAI that are similar to terms found in ASTM. An example is shown in **Fig. 2.3**. AAI uses the language "conditions indicative of a release or threatened release" in place of the ASTM term "recognized environmental condition" (REC).

These differences in terminology can be found throughout the documents. They should not be interpreted as subtle differences in meaning. On the contrary, great effort was taken to try to convey the same thought using different words.

2.3 Terminology of E1527

Section 3 of **E1527** consists of two subsections: Definitions (Section 3.2) and a List of Acronyms (Section 3.3). When reviewing the standard, readers will note that many of the terms are italicized. Italicized words in **E1527** alert the reader to the fact that there is a definition associated with that word or phrase.

In the first group of definitions, over 100 terms are defined. Terms such as "actual knowledge," "appropriate inquiry," "due diligence," "material threat," "obvious," and "practically reviewable" are given meaning within the context of the standard. Those of you familiar with previous versions of the standard will note that there used to be two sets of definitions. One set was common terms in general use, and one set was terms specific to **E1527**. Over the years, these distinctions have blurred and all terms were combined into one section in the 2005 version and remain so in the 2013 version.

FIG. 2.4   Previous definition of an HREC.

> Previous definition of *historical recognized environmental condition*—an environmental condition that in the past would have been considered a *recognized environmental condition*, but that may or may not currently be considered a *recognized environmental condition*. The final decision rests with the *environmental professional* and will be influenced by the current impact of the *historical recognized environmental condition* on the *property*. If a past release of any *hazardous substances* or *petroleum products* has occurred in connection with the *property* and has been remediated, with such remediation accepted by the responsible regulatory agency (for example, as evidenced by the issuance of a no further action letter or equivalent), this condition shall be considered an *historical recognized environmental condition* and included in the findings section of the *Phase I Environmental Site Assessment* report. The *environmental professional* shall provide an opinion of the current impact on the *property* of this *historical recognized environmental condition* in the opinion section of the *report*. If this *historical recognized environmental condition* is determined to be a *recognized environmental condition* at the time the *Phase I Environmental Site Assessment* is conducted, the condition shall be identified as such and listed in the conclusions section of the *report*.

FIG. 2.3   Comparison of terminology.

> AAI terminology: "conditions indicative of a release or threatened release"
>
> ASTM terminology: the presence or likely presence of any *hazardous substances* or *petroleum products* in, on, or at a *property*: (*1*) due to release to the environment; (*2*) under conditions indicative of a *release* to the *environment*; or (*3*) under conditions that pose a *material threat* of a future *release* to the *environment*.

FIG. 2.5 E1527-13 HREC definition.

> E1527-13 definition of *historical recognized environmental condition*—a past *release* of any *hazardous substances* or *petroleum products* that has occurred in connection with the *property* and has been addressed to the satisfaction of the applicable regulatory authority or meeting unrestricted use criteria established by a regulatory authority, without subjecting the *property* to any required controls (for example, *property* use restrictions, *activity and use limitations*, *institutional controls*, or *engineering controls*). Before calling the past *release* a *historical recognized environmental condition*, the *environmental professional* must determine whether the past *release* is a *recognized environmental condition* at the time the *Phase I Environmental Site Assessment* is conducted (for example, if there has been a change in the regulatory criteria). If the EP considers the past *release* to be a *recognized environmental condition* at the time the Phase I ESA is conducted, the condition shall be included in the conclusions section of the report as a *recognized environmental condition*.

FIG. 2.6 E1527-13 definition of CREC.

> E1527-13 definition of *controlled recognized environmental condition*—a *recognized environmental condition* resulting from a past *release* of *hazardous substances* or *petroleum products* that has been addressed to the satisfaction of the applicable regulatory authority (for example, as evidenced by the issuance of a no further action letter or equivalent, or meeting risk-based criteria established by regulatory authority), with *hazardous substances* or *petroleum products* allowed to remain in place subject to the implementation of required controls (for example, *property* use restrictions, *activity and use limitations*, *institutional controls*, or *engineering controls*). (See [**2**].) A condition considered by the *environmental professional* to be a *controlled recognized environmental condition* shall be listed in the findings section of the *Phase I Environmental Site Assessment report* and as a *recognized environmental condition* in the conclusions section of the *Phase I Environmental Site Assessment report*. (See [**3**].)

As we go through the various sections of the standard, we will revisit many of the definitions and demonstrate how they are used in practical applications of **E1527**.

Section 3.3 of the **E1527** provides explanations for the 27 acronyms used in the standard. Many of these are related to EPA regulatory programs and are common to the environmental industry.

One key revision in the **E1527-13** was the modification of the "historical recognized environmental condition" (HREC). Recall that a REC is tied to a "release"; therefore, any release that had occurred, even if it had been cleaned up, would technically qualify as a REC. The HREC was initially developed in the 2000 version of the **E1527** as a mechanism to allow past releases that had been cleaned up to no longer be identified as RECs. The previous HREC definition is shown in **Fig. 2.4**.

The definition was a bit cumbersome. Application of the HREC was further complicated by the increasing use of risk-based closures in which contaminants were allowed to remain in place given certain land use restrictions or engineering control requirements.

Recall from **Chapter 1** that the 2002 Brownfields Bill established the Bona Fide Prospective Purchaser (BFPP) Landowner Liability Protection that allows a buyer to purchase property with knowledge that the site is contaminated. However, along with this new ability to buy contaminated property with knowledge, there came "continuing obligations" that take effect upon acquisition of the property. If the Phase I ESA identified that a past release had occurred but had received regulatory closure, then it was not uncommon for the EP to identify this situation as an HREC without giving any consideration to the conditions associated with the closure.

The **E1527-13** revision addresses this challenge by splitting the HREC into two very specific types of closures. The new HREC definition (presented in **Fig. 2.5**) only applies to those past releases that have been addressed to unrestricted use. It is important to note that the HREC, as in previous versions, is not a REC. In other words, a release has occurred but, in essence, nothing remains.

The ASTM task group then worked to develop a new definition that would capture those past releases that had been addressed, but to a level that was only appropriate for a defined set of circumstances. This new definition is the "controlled recognized environmental condition" (CREC). A very important distinction between the HREC and the CREC is that the CREC must be included in the Conclusions section of the report. In effect, the CREC is a subset of a REC because of the "presence" of the contaminant. The CREC is defined in **Fig. 2.6**, and examples of common CREC conditions are presented in **Fig. 2.7**.

There are two very important "Notes" included in the CREC definition to assist in understanding the intent of the CREC. One note provides a very common example of a leaking underground storage tank site that has been addressed to commercial use standards. Because the cleanup does not meet unrestricted residential cleanup criteria, this would be an example of a CREC, with the "control" represented by the restriction that the property use remain commercial.

The second note makes it clear that a condition identified as a CREC does not mean that the EP has evaluated or confirmed the adequacy, implementation, or continued effectiveness of the control.

Another key definition to the evaluation of RECs is the concept of de minimis. **Figure 2.8** presents the ASTM **E1527** de minimis definition. Prior to the publication of **E1527-13**, "de minimis" was imbedded within the REC definition, but it is found in the **E1527-13** version as a stand-alone definition. De minimis is intended to apply to a situation in which a release has occurred (and thus would technically be a REC) but, based on the judgment of the EP, it would be insignificant. This would typically apply in the case of a small spill that is observed during a site visit. A condition determined by the EP to be de minimis is not a REC.

FIG. 2.7 A gas station with petroleum impacts meeting commercial cleanup criteria or developments constructed over approved hazardous waste repositories could be examples of a CREC.

FIG. 2.8 **E1527-13** definition of de minimis condition.

> 3.2.22 *de minimis condition*—a condition that generally does not present a threat to human health or the *environment* and that generally would not be the subject of an enforcement action if brought to the attention of appropriate governmental agencies. Conditions determined to be *de minimis conditions* are not *recognized environmental conditions* nor *controlled recognized environmental conditions*.

2.4 Looking at Sources of Related Terminology

In the early 1990s, **E1527** was initially developed concurrently with **E1528**—Transaction Screen [2]. The users on the E50 committee felt that many commercial transactions involved properties that did not require the level of inquiry prescribed by the Phase I ESA standard to meet the requirements of AAI. They felt that there was a need in the industry for a standardized practice that could be performed by a "nonenvironmental professional." The Transaction Screen was designed for just that purpose. It contained a terminology section identical to earlier versions of **E1527**. Since the publication of the EPA AAI rules, the Transaction Screen no longer satisfies the requirements for AAI because it is not as rigorous as what is required by AAI. The Transaction Screen has been revised and now can be used to conduct limited environmental due diligence screening, but it should not be used to satisfy AAI. The purpose of the revised standard is shown in **Fig. 2.9**.

Another document, ASTM **E2091** Standard Guide for Use of Activity and Use Limitations [3], has relevance to the Phase I process. It provides a discussion of general terminology including institutional and engineering controls.

On occasion, the user knows or has a strong suspicion that the subject property is contaminated. Phase I ESAs are sometimes conducted for the purpose of identifying RECs in full anticipation that further inquiry will be conducted in a Phase II portion of the project. The **E1903** Standard [4] and related standards associated with the Phase II investigations contain numerous applicable terminology sections. ASTM published a separate publication titled "ASTM Standards on Environmental Site Assessment Process Plus" [5]. This is an electronically available subscription to a compilation of standards. The individual standards contain terminology

FIG. 2.9 Purpose of **E1528-14** Transaction Screen.

> *Purpose*—The purpose of this practice is to define a good practice in the United States of America for conducting a *transaction screen* [1] for a parcel of *commercial real estate* where the *user* wishes to conduct limited environmental *due diligence* (that is, less than a *Phase I Environmental Site Assessment*). If the driving force behind the environmental *due diligence* is a desire to qualify for one of the *Comprehensive Environmental Response, Compensation, and Liability Act (CERCLA) Landowner Liability Protections (LLPs)*, this practice should not be applied. Instead, the ASTM Standard Practice **E1527** for Environmental Site Assessments: Phase I Environmental Site Assessment Process or ASTM **E2247**: Standard Practice for Environmental Site Assessments: Phase I Environmental Site Assessment Process for Forestland or Rural Property may be used.

sections; ASTM **D653** [6] is one of the standards included in this publication and contains 32 pages of ESA terminology. The Terminology Standard was developed jointly by the ASTM **D18** Committee on Soil and Rock and the Committee on Definitions and Standards of the Geo-technical Engineering Division of the American Society of Civil Engineers (ASCE). The two groups function together as the Joint ASCE/ASTM Committee on Nomenclature in Soil and Rock Mechanics.

2.5 Summary

In this chapter, we learned the importance of using the terminology defined by the standards. We learned that the terms defined in the Terminology section of the standard are an integral part of the Phase I ESA and are critical to understanding **E1527**. We demonstrated the importance of appropriate terminology in the practical application of **E1527**. We identified additional ASTM sources of information and terminology associated with the Phase I ESA.

References

[1] ASTM Standard **E1527**, Standard Practice for Environmental Site Assessments: Phase I Environmental Site Assessment Process, *Annual Book of ASTM Standards*, ASTM International, West Conshohocken, PA, 2013.

[2] ASTM Standard **E1528**, Standard Practice for Limited Environmental Due Diligence: Transaction Screen Process, *Annual Book of ASTM Standards*, ASTM International, West Conshohocken, PA, 2014.

[3] ASTM Standard **E2091**, Standard Guide for Use of Activity and Use Limitations, Including Institutional and Engineering Controls, *Annual Book of ASTM Standards*, ASTM International, West Conshohocken, PA, 2011.

[4] ASTM Standard **E1903**, Standard Practice for Environmental Site Assessments: Phase II Environmental Site Assessment Process, *Annual Book of ASTM Standards*, ASTM International, West Conshohocken, PA, 2011.

[5] ASTM Standards on Environmental Site Assessment Process, ASTM International, West Conshohocken, PA, 2013.

[6] ASTM Standard **D653**, Terminology Relating to Soil Rock and Contained Fluids, *Annual Book of ASTM Standards*, ASTM International, West Conshohocken, PA, 2011.

Chapter 3 | Significance and Use of E1527

3.1 Practical Application of E1527

The ASTM Standard **E1527** [1] was originally developed to satisfy one of the requirements of the Innocent Landowner Defense (ILD) under the Comprehensive Environmental Response, Compensation, and Liability Act (CERCLA). It was modified in 2005 to incorporate the new Landowner Liability Protections (LLPs) and further revised in 2013.

People may use the standard to comply with AAI to establish LLPs. It is also commonly used to help make business decisions regarding environmental impacts that might affect the property. The U.S. Environmental Protection Agency (EPA) requires grantees participating in the Brownfields program to follow the EPA All Appropriate Inquiries Final Rule (AAI). It is interesting to note that the ASTM **E1527** Standard is used in other countries, although these applications have nothing to do with LLPs because CERCLA and the related liabilities are a uniquely U.S. law. The fact that it is used outside of the legal context for which it was developed is an indication of its wider value in identifying environmental issues.

During the 1997 revisions of **E1527** [1], it became apparent that the needs of the user may extend beyond the CERCLA considerations. In order not to limit the application of the standard, the committee voted to include language regarding business risk that recognizes significantly broader application of this practice. The broader application of the standard is discussed in Sections 4.1 and 4.2 of **E1527**. In this chapter, we will review some specific uses and applications that demonstrate the practical impact of these sections.

3.1.1 VOLUNTARY USE

Before we begin discussing specific applications of **E1527**, it is important for the environmental professional (EP) to understand that the use of the standard is voluntary. To this day, many commercial real estate transactions are concluded without the Phase I Environmental Site Assessment (ESA) being performed. Three common reasons are as follows.

First, a buyer with financial resources to pay cash is not required by a lender to perform a Phase I as a loan condition. The buyer's risk tolerance may be such that he or she believes that the property is not sufficiently affected by contamination to cause substantial financial hardship. The buyer may have also taken different steps to limit liability such as other investigations or legal indemnifications.

Second, the use of the property or the loan value is often used by lenders to decide what level of inquiry to use. The lender may not require Phase I ESA on commercial residential use; apartment use may dictate a Transaction Screen or some other less involved due diligence. It is also common practice by lenders to select loan limit values that trigger different levels of inquiry. For example, loans under $250,000 may require no Phase I ESA, but any loan more than $250,000 may require a Phase I ESA. The risk of a less than $250,000 loan turning into a $500,000 or more liability is not sufficiently high for the loan institutions to change the approach.

Environmental scientists often fail to understand the reasoning behind such "risky" decisions. EPs look at the condition of the property, focusing on contaminants, associated user liability, and the potential EP's liability. Because EPs have seen numerous examples in which the cleanup cost exceeded the purchase price of the property, the cost of the property is not as significant. EPs' training and appreciation of contamination issues tend to make some EPs believe that choosing anything less than a Phase I ESA is irresponsible. On the other hand, the users see the condition of the property from a business decision point of view. The business decision is based on a risk tolerance that is generally different from that of EPs. The voluntary nature of the standard allows the users to incorporate their risk tolerance into the business decision. Risk tolerance is not necessarily another word for luck, although luck may be a component. A significant portion of user risk tolerance may come from the probability of environmental problems and past experience. Statistically, how often did a $50,000 loan turn into a $500,000 liability? Did it happen to the user or someone in the geographic area close to the site? Low frequency of occurrence may be perceived as an acceptable risk.

Third, the user may decide that the Phase I ESA performed to the **E1527** is not appropriate for the particular property and choose a different scope of work. They may be concerned with an entirely different set of issues not specifically covered by **E1527**, such as radon, lead, asbestos, or indoor air quality.

3.1.2 USER MAY NOT BE SEEKING LLPS

A user may not be concerned about exerting an LLP defense, or even know that he or she should be. The user may be using the Phase I ESA as a commercially prudent and reasonable inquiry to address business risk concerns. The Phase I ESA can be applied as the first step in addressing known or suspected contamination issues. The contamination issues may be limited to CERCLA and petroleum products (which under certain circumstances are included in the scope of the E1527), or they may include other contaminants such as asbestos-containing building materials and lead-based paint (which are beyond the E1527 scope). It is imperative for the EP to be cognizant of the user's intent. The user, with the assistance of the EP, may elect to perform a more comprehensive inquiry by expanding the scope or limit the scope to address specific business risk concerns.

3.1.3 DESIGNED TO IDENTIFY RECOGNIZED ENVIRONMENTAL CONDITIONS

The Brownfields Amendment and AAI include broad-based principles aimed at obtaining reasonably ascertainable, commonly known, or other information that could be determined by inspection and research. The way this actually occurs in the ASTM Standard is that a prescriptive set of actions is laid out to help identify recognized environmental conditions (RECs). "Recognized environmental conditions" is not a term found in CERCLA, but it is consistent with the AAI objective of "identifying conditions indicative of releases or threatened releases," and finding RECs is the objective of the ASTM Phase I process. The ASTM Standard is structured in such a way that many findings do not rise to the level of a REC, and many RECs would not be significant enough to create a liability under CERCLA. The same can be said for the more general objective stated in the AAI rule. In other words, the process is very conservative in that it tends to identify issues well below the threshold that would involve CERCLA liability. The ASTM definition of a REC was modified in the E1527-13 Standard to align more closely with the AAI objectives while still providing valuable guidance for the EP. The revised REC definition is shown in **Fig. 3.1**.

The first priority for the EP is to identify RECs as they are defined above. Because the definition must be applied to a specific unique situation, EPs may struggle with what to identify as a REC in the report. Likewise, many users struggle with large lists of RECs and prefer to see reports with no RECs.

In the 2000 version of the ASTM Standard, a definition of "material threat" was added to help clarify the ASTM definition of a REC. There is little disagreement about what constitutes a REC when it involves a past release or a present release. Major disagreement is usually related to the REC concept of material threat of a future release. It was for this reason that material threat was defined in the 2000 revision. The material threat definition is provided in **Fig. 3.2**.

The presence of a hazardous substance or petroleum product when there is no indication of a past or present release forces a determination by the EP with respect to whether the conditions indicate a material threat of a future release. Although reasonable people differ on what constitutes a material threat, EPs need to explain the logic involved in their determination.

The other area of disagreement among EPs is the concept of "likely presence." Because a REC is the presence or likely presence, there is more judgment required in a determination of likely presence. As discussed above, the concept of presence is usually straightforward—it is observed or known to exist. Likely presence exists when, in the opinion of the EP, there is a high probability that a hazardous substance is present. There are two situations when this might occur. First is a situation in which there is a documented release with limited data available on extent or dispersal. A judgment has to be made about whether there is a likely presence. An example of the first situation is an offsite release that is upgradient and could result in contamination entering the property as a groundwater or vapor plume. The second common situation is when the past history of land use (old gas station, dry cleaner) suggests that releases probably occurred. In this second case, the EP is opining that the historical operating practices are enough to indicate a likely presence although there may be no observed or as-yet documented contamination.

The Report section of the ASTM Standard was revised in 2000, 2005, and again in 2013 to make it clear that the EP must do more than provide a list of RECs. The EP has some latitude in what they will call a REC. It is important to the user to understand the logic and reasoning of the EP. The ASTM Standard is very specific in requiring the EP to list all of the findings and then explain why they are or are not considered to be a REC. If not actually observed,

---

FIG. 3.1 The **E1527-13** definition of a REC.

> 1.1.1 Recognized Environmental Conditions—In defining a standard of good commercial and customary practice for conducting an environmental site assessment of a parcel of property, the goal of the processes established by this practice is to identify recognized environmental conditions. The term "recognized environmental conditions" means the presence or likely presence of any hazardous substances or petroleum products in, on, or at a property (1) due to a release to the environment, (2) under conditions indicative of a release to the environment, or (3) under conditions that pose a material threat of a future release to the environment. De minimis conditions are not recognized environmental conditions.

FIG. 3.2 Material threat definition from the **E1527-13** Standard.

> 3.2.52 *material threat*—a physically observable or *obvious* threat which is reasonably likely to lead to a release that, in the opinion of the *environmental professional*, is threatening and might result in impact to public health or the environment. An example might include an aboveground storage tank system that contains a *hazardous substance* and which shows evidence of damage. The damage would represent a *material threat* if it is deemed serious enough that it may cause or contribute to tank integrity failure with a release of contents to the environment.

this REC opinion discussion is focused on explaining if there is a material threat or if there is a likely presence. Additional options to define concerns came with the 2013 revision to the Standard including controlled recognized environmental conditions (CRECs) and a revised definition of historical recognized environmental conditions (HRECs).

Figure 3.3 depicts chemical storage in 55-gal steel drums. Let us assume that while conducting the Phase I ESA, the EP noted this area. The nature of the chemical is such that it could cause significant contamination to the property if it was spilled. None of the drums appear to be leaking. Some of the drums have some rust, but they are not cracked or broken; they are sealed and appropriately labeled. The current owner informs the EP during the interview that the drums will be removed from the property. The EP takes a photograph of the drum storage and documents the content of the interview in the report. The EP decides not to list this as a REC, reasoning that the condition does not constitute a material threat of a release, and explains his or her reasoning in the report. The EP might reason that there was no evidence of spills, the drums were located in a protected area of the facility, and the floor was concrete with no signs of cracking. The simple presence does not constitute a material threat of a future release.

Now let us consider the identical scenario with one difference. The EP lists the drums as a REC because he or she observes that there is no secondary containment, so that if the chemical was to leak it could result in an environmental impact on the subject property. In the opinion portion of the report, the EP describes the condition of the drums (some rusted), provides the user with photographs of the drums, and opines that considering all of the other information in the report (condition of drums, poor management practices, vague information of operational practices), this is a

FIG. 3.3   Chemical storage in drums.

Source: Provided by J. Kilgore.

REC on the basis of the opinion that it constituted a material threat of a future release. Although it did not appear to adversely affect the subject property, a release could have a significant effect because of shallow groundwater under the facility.

The day after the EP visited the site, while removing the chemical drums from the property to enable the buyer to move his operation in, the forklift brakes fail and the forklift punctures several of the drums. The chemical is washed off of the pad with water and contaminates the property. A month after the sale the contamination migrates onto the neighboring property and kills the landscaping. The neighbor investigates and finds out about the chemical that killed his trees. He calls the state agency. One thing leads to another and the remediation action is going to cost more than the loan value. The user, now the owner of the liability, tries to find someone to blame, reads the **E1527** definition of a REC for the first time, and calls the EP:

"Did you know about the drums? How come they were not listed as a REC?"

The EP answers:

"I documented the drums in the report, and the pictures are in the file!"

Both EPs documented the condition observed at the site in their reports. The only difference was that one interpreted the condition as a REC whereas the other EP listed it as a finding. The scenario is based on a true story, and the report could have been written either way. There are several valid reasons why some conditions such as the one described in the above scenario are not listed by many EPs as a REC.

Regardless of the interpretation, EPs must document the conditions of the subject property in the report. If they are not listed as RECs, then they must be listed as findings and have an opinion provided as to their significance.

The presence of RECs on the property does not mean that the property is contaminated. RECs describe the conditions of the property with the potential of causing, or having caused, contamination of the subject property. The information enables users to develop a risk perception about the property. It is up to the user, not the EP, to decide what to do about the RECs. The EP's function is to provide sufficient information necessary to assist the user in reaching that decision.

3.1.4 RELATIONSHIP TO E1528

The **E1528** Standard [2] is commonly known as the Transaction Screen (TS). This document was initially designed to meet all of the requirements of "all appropriate inquiries" for an ILD in which a user deems the environmental risks on a property to be very low and would not require the expertise of an EP. The passage of the Brownfields Amendment brought into question the efficacy of the TS. The 2002 Brownfields Bill cited only the **E1527** as the interim standard for conducting all appropriate inquiries until EPA promulgated its AAI regulation, and EPA's November 2005 AAI was very specific in saying that ASTM **E1527-13** met the requirements. Neither EPA nor Congress have indicated that the TS met the requirements of AAI.

In response, the TS was revised to remove all references to addressing issues related to AAI and LLPs. It has been reissued to

FIG. 3.4 Ten criteria for AAI set forth in the 2002 Brownfields Amendments to CERCLA.

> The criteria are set forth in CERCLA section 101(35)(2)(B)(iii) and include:
> - The results of an inquiry by an environmental professional;
> - Interviews with past and present owners, operators, and occupants of the facility for the purpose of gathering information regarding the potential for contamination at the facility;
> - Reviews of historical sources, such as chain of title documents, aerial photographs, building department records, and land use records, to determine previous uses and occupancies of the real property since the property was first developed;
> - Searches for recorded environmental cleanup liens against the facility that are filed under federal, state, or local law;
> - Reviews of federal, state, and local government records, waste disposal records, underground storage tank records, and hazardous waste handling, generation, treatment, disposal, and spill records, concerning contamination at or near the facility;
> - Visual inspections of the facility and of adjoining properties;
> - Specialized knowledge or experience on the part of the defendant;
> - The relationship of the purchase price to the value of the property, if the property was not contaminated;
> - Commonly known or reasonably ascertainable information about the property;
> - The degree of obviousness of the presence or likely presence of contamination at the property, and the ability to detect the contamination by appropriate investigation.

serve as a screening document for those interested in identifying environmental issues but not desiring to comply with AAI.

The TS is often used as a preliminary information-gathering tool about properties. Using that information, the users then select candidates for Phase I, Phase II, or any other investigation they may deem appropriate. A TS might also be an appropriate tool in the case of a refinance in which an underlying Phase I ESA has already been conducted and there is a desire to obtain updated information regarding any new potential environmental conditions affecting the site.

The ATS is also used on less expensive properties where the users feel that they cannot justify spending the money for a full Phase I. The use of a TS by lenders has increased since the passage of the Secured Creditors Exemption in 1996, which potentially limits the lenders' CERCLA liability.

3.2 The Principles Behind AAI

In the Brownfields Amendment, Congress specified ten criteria to be included in the regulatory standards and practices to be established by EPA. These are provided in **Fig. 3.4**.

The first six of these were clearly intended to reflect the ASTM **E1527** Standard Practice. The last four are a restatement of language found in an earlier CERCLA legislation and are sometimes referred to as the "SARA artifacts," a reference to a series of factors set forth in the 1986 CERCLA amendments that Congress instructed a court to consider (SARA = Superfund Amendments and Reauthorization Act).

The AAI was developed by EPA to conform to the criteria listed in **Fig. 3.4**. Before describing the specific steps to be followed, EPA cites objectives, which take the tasks outlined above and describe the information that is to be obtained. The specific language found in the Final Rule is in **Fig. 3.5**.

The following observations about the above language are of interest. The first, the somewhat obscure reference to persons identified under § 312.1(b)(1), is EPA language for anyone seeking to establish one of the LLPs. In ASTM terminology, we would simply call them "the user." Note that when we substitute "user," the language says "… the user and the EP, must seek …". This is a good example of two problems with the Brownfields Amendment and the AAI. They frequently use that common introduction to a statement about something that must be done.

FIG. 3.5 Objectives of AAI.

> (1) In performing the all appropriate inquiries, as defined in this section and provided in the standards and practices set forth in this subpart, the persons identified under § 312.1(b)(1) and the environmental professional, as defined in § 312.10, must seek to identify through the conduct of the standards and practices set forth in this subpart, the following types of information about the subject property:
>   (i) Current and past property uses and occupancies;
>   (ii) Current and past uses of hazardous substances;
>   (iii) Waste management and disposal activities that could have caused releases or threatened releases of hazardous substances;
>   (iv) Current and past corrective actions and response activities undertaken to address past and on-going releases of hazardous substances;
>   (v) Engineering controls;
>   (vi) Institutional controls; and
>   (vii) Properties adjoining or located nearby the subject property that have environmental conditions that could have resulted in conditions indicative of releases or threatened releases of hazardous substances to the subject property.

It is not very clear who is supposed to do it: the user or the EP or both.

A very similar paragraph is included in AAI to deal with Brownfields grantees. The only difference is that it adds petroleum products and controlled substances after the term "hazardous substances" in paragraphs (ii), (iii), and (vii). The reason for this is not intuitively obvious. It can be traced to the fact that petroleum products are specifically exempt from CERCLA and are regulated under other environmental laws, but they are of great interest in the context of Brownfields grants. The inclusion of controlled substances for the purposes of Brownfields grants was an effort to provide funding to assist grantees with the very common problem of clandestine drug laboratories, although controlled substances (methamphetamine or otherwise) are not as listed CERCLA hazardous substances. Congress wanted to allow grant funds to be used for assessing Brownfields sites that had problems related to petroleum products or clandestine drug laboratory use. They could include it in the grant portion of the Brownfields Amendment, but they could not incorporate it in the traditional CERCLA liability portions of the statute. The AAI only requires evaluation of petroleum products or controlled substances when associated with Brownfields grants. They are not mentioned in the context of LLPs because they are not CERCLA issues; therefore, they could not lead to a Superfund liability that would be subject to an LLP defense.

3.3 The Principles Behind E1527

When performing a Phase I ESA, the EP can always look deeper or gather some more information. It is often a difficult task to determine where to draw the line, which tasks should be included, or how much information should be gathered. **E1527** recognizes the validity of these issues and provides the EP with some guidance. The guidance is based on some of the fundamental principles of the **E1527** Standard.

First and foremost, performing a Phase I ESA, or any other environmental assessment or investigation, does not eliminate uncertainty and, at best, can only reduce it to an acceptable level for the user. There are no guarantees.

The Phase I ESA process gathers only existing information; it does not create it. EPs are not doing any testing or engaging in any activity that generates new information. The EPs can be looked upon as environmental information and knowledge brokers.

The process is not an exhaustive search and generally has to be accomplished within specified time limits. These are usually dictated by users' needs. Time constraints are real, limit the ability to gather certain information, and should be noted as limitations to the ESA.

The Phase I ESA is a snapshot of the property at a specific point in time. This is an important distinction from an audit, which looks at the continuing activities associated with the property. Consequently, if the environmental assessment is evaluated at a later date by other parties, the evaluators need to take a step back in time, considering the level of knowledge and state of the art at the point in time when the Phase I ESA was conducted. Once the EP leaves the property, the conditions can immediately

FIG. 3.6 **E1527** language from Section 4.5.4.

> Subsequent environmental site assessments should not be considered valid standards to judge the appropriateness of any prior assessment based on hindsight, new information, use of developing technology or analytical techniques.

change. The property may become contaminated by a truck that pulls onto the subject property an hour after the assessment and dumps 4,000 gal of oil contaminated with polychlorinated biphenyls. Likewise, what constitutes a REC is a moving target that changes with time. Spraying used motor oil on roads to reduce dust was an acceptable practice in the past and was considered as having a de minimis environmental impact on the property. Today, this practice would constitute a REC, and the impacts on the subject property can be far from de minimis. **Figure 3.6** is an excerpt from Section 4.5.4 of the standard that recognizes the constantly changing nature of the industry and environmental laws.

3.4 The Shelf Life of Reports

Considering the nature of potentially rapid change in the conditions of the subject property, the shelf life of a report could be perceived as extremely short. The Phase I ESA report could literally be considered obsolete by the time it is written. Recognizing the problem, the standard establishes some ground rules for the continued viability of the Phase I ESAs. Section 4.6 in **E1527** lists specific rules. The AAI resulted in some changes to the shelf life of the reports from the previous editions of the Phase I standard. The two principal changes included a more specific definition of the dates involved and a statement that the reports expire after 1 year. These changes are incorporated in the 2005 and 2013 versions of the ASTM Standard.

Note that AAI also makes specific reference to the time as being measured from the time the work was conducted. This is different from the date on the report. For jobs done with a 1-week turnaround, it may be of little practical impact, but it can be significant for a Phase I ESA that had a 30-day turnaround or instances in which a report is not in final form for several weeks or months. For example, if the site visit was done 30 days before the report date, the effective useful life of that component of the report is not 180 days but 150 days. For this reason, it is important to document when the various Phase I ESA activities occurred.

The dates are specific to the actual date of acquisition of the property, or for transactions not involving acquisitions the date of the intended transaction. This becomes important in the context of commercial real estate because it is not uncommon to have multiple parties involved and find that a closing is held up until some final detail is completed. In some situations the issues related to the Phase I ESA may all be resolved to everyone's satisfaction, but the transfer cannot occur until some other loan document is obtained. If the closing slips past the 180-day or 1-year time deadlines, the

FIG. 3.7 Requirements for updates after 180 days.

> (i) *Interviews* with *owners*, *operators*, and *occupants*;
> (ii) Searches for recorded environmental cleanup liens;
> (iii) Reviews of federal, tribal, state, and local government records;
> (iv) Visual inspections of the *property* and of *adjoining properties*; and
> (v) The declaration by the *environmental professional* responsible for the assessment or update.

Source: ASTM **E1527-13**.

Phase I ESA, or a portion of the Phase I ESA, may have to be revisited to maintain AAI compliance.

Broadly speaking, Phase I ESAs are considered valid for 6 months (180 days). After this time, the use of the ESA is subject to specific rules. Perhaps the most problematic of all is the rule requiring the party relying on the past Phase I ESA to determine whether it met or exceeded the requirements of **E1527**. Now the EPs' or users' function suddenly changes from an information user to a judge of other EPs' work product. From a practical standpoint, this function could be a difficult and potentially litigation-prone task. The EPs or users relying on previous Phase I ESAs need to be very familiar with this section of the standard and use appropriate professional judgment.

After 180 days, the Phase I ESA report (that met or exceeded the requirements of ASTM **E1527**) can be updated. The requirements for the update are listed in **Fig. 3.7**.

For reports older than 1 year, the report cannot just be updated but must be redone. The standard goes on to say that it is possible to use information from a prior report if it was generated using procedures meeting or exceeding the requirements of ASTM **E1527**. It also says that the information cannot be used without current investigation of conditions likely to affect RECs.

This raises the question—Is there any difference? As a practical matter—not much. In both cases, the principal part of the report that is salvageable is the historical review and the general site description. Just about everything else has to be redone. In the 180-day update an EP could attach a document that was titled "Update" and discusses the information gathered in the update. After the 1-year expiration, the EP would do essentially the same work and provide a new report that incorporated the usable information from the expired report.

For liability reasons, some consultants are reluctant to use any information that they did not specifically obtain. This is especially true if they are not familiar with the firm or person who did the prior work. In using an old report without independent verification of the data, you are assuming that whoever did the work performed it in a competent manner.

There is one other interesting change in the wording related to AAI and the use of a Phase I ESA report. AAI and ASTM both acknowledge that other people besides the intended user may end up "using" the report. The standard goes on to say that any subsequent users other than one for whom the report was prepared must satisfy the users' responsibilities as described in various sections of the standard and reflected in Appendix X-3. This is discussed in more detail in **Chapter 6**.

3.5 Rules of Engagement

It is up to the EP and the user to work out the contractual and legal obligations associated with Phase I ESAs. As we discussed in this chapter, Section 4 of **E1527** provides some guidance with respect to the significance and use of the standard, but the contractual issues are specifically excluded and have to be worked out by the EP and the user. The rules of engagement may vary depending on the relationship of the players, user needs, EPs' operational protocols, business judgment, and other issues.

Whenever the applications of the standard go beyond the scope of the standard, the users and EPs must pay particular attention to contracting to ensure that the additional items and responsibilities are sufficiently addressed in the contract.

3.6 Summary

In this chapter, we learned about varied applications of the standard, some of which extend beyond the scope of CERCLA. We discussed how the user's needs drive the application of the Standard. We reviewed the relationship of the TS, AAI, and Phase I ESAs. We talked about the principles driving the application of the Phase I standard, noting the limited shelf life of the reports. We learned that contractual issues are not included in the scope of **E1527**.

References

[1] ASTM Standard **E1527**, Standard Practice for Environmental Site Assessments: Phase I Environmental Site Assessment Process, *Annual Book of ASTM Standards*, ASTM International, West Conshohocken, PA, 2013.

[2] ASTM Standard **E1528**, Standard Practice for Limited Environmental Due Diligence: Transaction Screen Process, *Annual Book of ASTM Standards*, ASTM International, West Conshohocken, PA, 2014.

Chapter 4 | Responsibilities

4.1 Responsibilities Under AAI

The Brownfields Amendment and the U.S. Environmental Protection Agency (EPA) All Appropriate Inquiries Final Rule (AAI) [1] have significant implications for the users and the environmental professionals (EPs). As discussed in **Chapter 3**, the ten criteria included in the Brownfields Amendment provided guidance that EPA has to consider in developing the rules for AAI. Those criteria speak in terms of the user and the EP. It makes no specific distinction between the user and the EP. AAI does not actually make use of the term "user." It uses a much more involved definition to account for various parties involved in transactions, including grantees. The language from AAI on applicability may be found in **Fig. 4.1**. To simplify the concept, we will use the ASTM E1527 terminology for user [2].

The Brownfields Amendment is not very specific about the roles of the user and the EP. The AAI is somewhat more specific and details specific tasks required of the EP. The ASTM Standard provides more clarification of the roles of the parties involved.

One area of concern is the users' responsibility, which can be divided into two general subjects: conducting AAI and continuing obligation requirements. This is a major change in the user responsibilities from pre-Brownfields Amendment practices.

The Phase I Environmental Site Assessment (ESA) identifies "legacy contamination" as known or suspected soil or groundwater impacts from activities conducted long ago. As we move forward in a more sophisticated environment of managing risks from past releases, we create what could be referred to as "legacy remedies." Because it is often not economically viable (or necessary) to remove all past contamination, as a society we accept and acknowledge that some contaminants can remain in place as long as they are correctly managed. One of the postacquisition continuing obligations is related to Activity and Use Limitations (AULs), which comprise the range of mechanisms used to manage these residual environmental impacts. Section 5 of the ASTM Standard discusses the significance of AULs. The user and the EP have responsibilities to identify AULs in the Phase I process. These could be institutional controls that take the form of land use restrictions or engineering controls that may include a physical cap or an operating remediation system (see **Fig. 4.2**). The user must identify and comply with these AULs as part of the Landowner Liability Protection (LLP) defenses. Contained within the preamble to the rule, EPA states that failure to discover the AUL in the Phase I process does not relieve the user from the responsibility to comply.

Unfortunately, there is no consistent place to find AULs. They are sometimes included in land title records. Some state environmental regulatory agencies maintain registries of AULs. In some cases, they are found in judicial records. For many actions, project case files may be the only place they are documented. Over time, it is likely that more robust systems will evolve for tracking AULs. Until that time, finding them will be highly specific to the site.

Another related term used in AAI and ASTM regarding encumbrances on the property is the environmental lien. An environmental lien would be a financial encumbrance filed by the environmental agency against the property to recover costs associated with addressing a release at the property (see **Fig. 4.3**). There are two types of environmental liens. One is a traditional lien, which is filed by EPA to recover response costs associated with a Comprehensive Environmental Response, Compensation, and Liability Act (CERCLA) response action. A relatively recent type of lien is the windfall lien that derives from the provision in the 2002 Brownfields Amendments that is specifically tied to a qualified Bona Fide Prospective Purchaser (BFPP) in which the government may recover costs they incur in cleaning up a site when the land sells up to an amount that does not exceed the increase in fair market value of the property that is directly attributable to the response action. The actual amount of recovery is a complicated determination

FIG. 4.1   AAI definition of "user."

> The requirements of this part are applicable to
> (1) Persons seeking to establish
>   (i) The Innocent Landowner Defense pursuant to CERCLA Sections 101(35) and 107(b)(3);
>   (ii) The Bona Fide Prospective Purchaser liability protection pursuant to CERCLA Sections 101(40) and 107(r);
>   (iii) The Contiguous Property Owner liability protection pursuant to CERCLA Section 107(q); and
> (2) Persons conducting site characterization and assessments with the use of a grant awarded under CERCLA Section 104(k)(2)(B).

Source: From AAI.

FIG. 4.2   Definitions of ECs and ICs.

> *Engineering controls (ECs):* Physical modifications to a site or facility (for example, capping, slurry walls, or point-of-use water treatment) to reduce or eliminate the potential for exposure to hazardous substances or petroleum products in the soil or groundwater on the property.
>
> *Institutional controls (ICs):* A legal or administrative restriction (for example, "deed restrictions," restrictive covenants, easements, or zoning) on the use of, or access to, a site or facility to (1) reduce or eliminate potential exposure to hazardous substances or petroleum products in the soil or groundwater on the property, or (2) to prevent activities that could interfere with the effectiveness of a response action to ensure maintenance of a condition of no significant risk to public health or the environment.

Source: From ASTM **E1527-13**.

FIG. 4.3   Definition of environmental lien.

> *Environmental lien:* A charge, security, or encumbrance upon title to a *property* to secure the payment of a cost, damage, debt, obligation, or duty arising out of response actions, cleanup, or other remediation of *hazardous substances* or *petroleum products* upon a *property*, including (but not limited to) liens imposed pursuant to CERCLA 42 U.S.C. §§9607(1) & 9607(r) and similar state or local laws.

Source: From ASTM **E1527-13**.

based on several factors, including cost of remediation, value of the land, and the increased value associated with the cleanup. Often these Windfall Liens are prenegotiated between a BFPP and the agency to eliminate the financial risk of a future unknown demand. Parallel state programs can also place liens on a property.

There are very few of these liens in existence, but they will increase with time and the continued emphasis on Brownfields redevelopment. These liens are very closely associated with BFPP-type projects. Environmental liens are typically found during the course of research for the issuance of title insurance and would be shown in the policy of title insurance issued to the owner, buyer, or lender. In limited parts of the country and where required by law, these liens are found in judicial records.

4.2   User Responsibilities ASTM

Members of the E50 Committee have been teaching the ESA standards since 1994. Students, many of them EPs, are often surprised to find out about the number of responsibilities the users have with respect to the Phase I ESA. The user responsibilities reflect the responsibilities set forth in CERCLA that are required on the part of a "defendant."

Users' responsibilities and the reasoning behind these user requirements are spelled out in Section 6 of **E1527** [1]. This section was in existence before EPA's AAI regulation, but it was not until the 2013 revision that the **E1527** articulated the federal mandates that set the basis for why these specific tasks are the responsibility of the user. Appendix X-3 of the **E1527-13** standard is an optional user questionnaire that could be used to assist in gathering this user-provided information. ASTM says it is mandatory for the user to provide the information to the EP whereas the AAI indicated that the user "should" provide the information but stopped short of requiring it. This was largely in response to comments received on the draft rule process in which users indicated that they desired confidentiality. EPA does state that if the information is not provided to the EP, then the EP is required to treat the lack of information as a data gap.

One of the issues for the EP is to determine who the user is. The user has typically been a single client or maybe the buyer and their lender. Part of Appendix X-3 includes a request for the client (user) to provide a list of other users of the document. If those other users wish to assert an LLP, then they must also document their participation in the process and should complete the questionnaire or at least document that they looked for the information. What responsibility the EP has, if any, to assist in this process is not clear. At a minimum, the EP should advise the client to forward the questionnaire to other users.

Appendix X-3 consists of two parts. There are six questions that specifically address the users' responsibilities outlined in AAI. The balance of the questionnaire is indicating the type of information that a user typically possesses that should be provided to the EP. Several of these items are typically covered in related documents such as the contract, proposal, or scope of work.

The ASTM Standard also has other site-specific information detailed in Section 10 that is related to helpful documents. Because this site-specific information is not outlined by statute, and could be obtained by others in addition to the user, these documents are not included in Appendix X-3. They are to be requested before the site visit. The standard requires that the property owner, key site manager, and user be asked if any of the documents can be provided.

Some of the responsibilities outlined in the standard are not well understood by the users. Not having read the standard, many users are not even aware of them. Sometimes the EPs do not communicate the responsibilities to the users, or users expect the EPs to take on these responsibilities.

The users are required by the standard to check title records (and judicial records where applicable) for environmental liens and AULs. The standard is very clear that the EP is not responsible for this activity unless the user specifically requests that the EP contract with a title professional.

If the user has any specialized knowledge or experience that would assist the EP in identifying recognized environmental conditions (RECs) on subject property, then they must report it to the EP. In addition, the standard requires the user to provide any helpful documents. Both of these are to be provided before the site reconnaissance.

These responsibilities often get muddled in the process because of multiple user issues. In general, the lender or buyer hires the EP to perform the Phase I ESA. As we discussed in **Chapter 1**, the lender and the buyer can be considered users by the **E1527**

Standard. The sophisticated lenders are sometimes aware of the requirement of the standard, but they generally possess very little specialized knowledge about potential RECs on the property, and lenders are not generally seeking federal LLPs. The buyers generally have more knowledge about the property, but more importantly, it is the buyers who are the intended "defendants" under the CERCLA framework.

Figure 4.4 shows one method used by the EPs to put users on notice by sending a written request for such information. It is important for the EPs to identify all of the users when making these information requests.

The owner of the property generally has most of the knowledge, but they may not be the "user" of the report. Unless the owner is also the user, the owner is under no obligation to provide any information and under certain circumstances may be attempting to conceal or pretend ignorance with respect to RECs. As we will discuss in **Chapter 8**, which addresses the interview process, the best manner in which the EPs can inquire about owners' knowledge and experience is through the interview process. A form, similar in content to **Fig. 4.4**, sent to the owner before an interview can also be used as a tool to put the owners or site managers on notice about some of the information that the EPs are seeking.

Note that this discussion about helpful documents is not specifically related to Appendix X-3, which is the documentation required by users to conform with AAI. Appendix X-3

FIG. 4.4 Request form for information.

**PHASE I
ENVIRONMENTAL SITE ASSESSMENT
USER-FURNISHED INFORMATION**

The following is a list of documents and information which could be useful to Avanti Environmental, Inc in preparing your Phase I Environmental Site Assessment. Check the appropriate boxes below, sign, and return this along with copies of any reasonably available documents or information. This will be attached to and made part of your completed Phase I Report.

Yes	No		
☐	☐	1.	Environmental site assessment reports
☐	☐	2.	Environmental compliance audit reports
☐	☐	3.	Environmental permits (i.e. solid waste disposal permits, hazardous waste disposal permits, wastewater permits, NPDES permits, underground injection permits)
☐	☐	4.	Registrations for underground and above-ground storage tanks
		5.	Registrations for underground injection systems
☐	☐	6.	Material safety data sheets
☐	☐	7.	Community right-to-know plan
☐	☐	8.	Safety plans: preparedness and prevention plans; spill prevention, countermeasure, and control plans; etc.
☐	☐	9.	Reports regarding hydrogeologic conditions on the property or surrounding area
☐	☐	10.	Notices or other correspondence from any government agency relating to past or existing environmental liens encumbering the property
☐	☐	11.	Geotechnical studies
		12.	Risk Assessments
		13.	Recorded AULs
☐	☐	14.	Information concerning any pending, threatened, or past litigation or administrative proceedings relevant to hazardous substances or petroleum products
☐	☐	15.	Notices from any governmental entity regarding any possible violation of environmental laws or possible liability relating to hazardous substances or petroleum products

The following are examples of information not specifically included in the helpful documents requirements as an example of the types of additional information that some EPs ask for as part of the process.

☐	☐	Disclosure of sumps, pits, drainage systems-existence and location
☐	☐	Building plans (architectural, utility, structural)
☐	☐	Description of current site operations, including layout drawings or sketches
☐	☐	Title report/chain of title
☐	☐	Tax assessor records (previous owner and occupants)
☐	☐	Purchase price analysis (if lower than comparable)
☐	☐	Current and historical photographs of site

I have reviewed the above list and have provided copies of documents and information that exists that could be obtained within reasonable time and cost constraints.

Signature _____ Date _____

includes a question about specialized knowledge. Specialized knowledge on the part of the user and helpful documents from the user, owner, and occupants are to be requested and provided before the site visit.

4.3 EP Responsibilities

Up until the 2005 version of the standard, the EP definition was intentionally vague and tied to sufficient training and experience to develop opinions regarding RECs. It contained no specific requirements for training, education, or experience.

This definition was one of the most controversial issues of the EPA AAI rule-making process. It was the subject of extensive comment after the draft rule was published. The final definition used in AAI is shown in **Fig. 4.5** and is incorporated directly into the ASTM Standard by reference in the Definition section and inclusion as Appendix X-2 of the standard.

One of the key concepts to all of the categories of the EP definition is the concept of relevant experience. This is also defined in AAI and is included in ASTM Appendix X-2 (see **Fig. 4.6**).

There is one more important distinction in the AAI Rule regarding the role of the EP. AAI introduces the concept of "responsible charge." This means that the EP need not conduct the actual work, but it must be done under his or her direct supervision. This is not a responsibility to be taken lightly, and there are generally accepted principles for what constitutes responsible charge.

4.4 Attributes of Responsible Charge

Neither ASTM nor the EPA AAI regulation define "responsible charge." The following points should be considered as guidelines for EPs to consider when determining whether they are in responsible charge of a Phase I ESA project:

- Supervise the individual performing the work
- Obtain or set the project parameters or criteria

FIG. 4.6 Definition of relevant experience.

Relevant experience, as used in the definition of environmental professional in this section, means participation in the performance of all appropriate inquiries, investigations, environmental site assessments, or other site investigations that may include environmental analyses, investigations, and remediation that involve the understanding of surface and subsurface environmental conditions and the processes used to evaluate these conditions and for which professional judgment was used to develop opinions regarding conditions indicative of releases or threatened releases (see §312.1(c)) to the subject property.

Source: From AAI.

FIG. 4.5 Definition of EP.

Environmental professional means

(1) A person who possesses sufficient specific education, training, and experience necessary to exercise professional judgment to develop opinions and conclusions regarding conditions indicative of releases or threatened releases (see § 312.1(c)) on, at, in, or to a property, sufficient to meet the objectives and performance factors in § 312.20(e) and (f).

(2) Such a person must
   i. Hold a current Professional Engineer's or Professional Geologist's license or registration from a state, tribe, or U.S. territory (or the Commonwealth of Puerto Rico) and have the equivalent of three (3) years of full-time relevant experience; or
   ii. Be licensed or certified by the federal government, a state, tribe, or U.S. territory (or the Commonwealth of Puerto Rico) to perform environmental inquiries as defined in § 312.21 and have the equivalent of three (3) years of full-time relevant experience; or
   iii. Have a Baccalaureate or higher degree from an accredited institution of higher education in a discipline of engineering or science and the equivalent of five (5) years of full-time relevant experience; or
   iv. Have the equivalent of ten (10) years of full-time relevant experience.

(3) An environmental professional should remain current in his or her field through participation in continuing education or other activities.

(4) The definition of environmental professional provided above does not preempt state professional licensing or registration requirements such as those for a professional geologist, engineer, or site remediation professional. Before commencing work, a person should determine the applicability of state professional licensing or registration laws to the activities to be undertaken as part of the inquiry identified in § 312.21(b).

(5) A person who does not qualify as an environmental professional under the foregoing definition may assist in the conduct of all appropriate inquiries in accordance with this part if such person is under the supervision or responsible charge of a person meeting the definition of an environmental professional provided above when conducting such activities.

Source: From AAI.

- Ability to require changes to the work and dictate the manner and methods by which the work is performed or the scope of involvement by the individual, even over the individual's objections, without negative impact on the EP's job status, advancement, or compensation
- Involved from start to finish as the individual performs the work
- Procedures in place for quality control and authority over the work that ensure that the EP is in control of the individual(s) performing the work
- Sufficient time spent with the individual during the course of the performance of the work sufficient to be familiar with the details of the work
- Familiar with the capabilities and methods of the individual performing the work
- Train the individual
- In close proximity or have readily accessible contact with the individual either by physical location or by frequent, clear, and full communication in verbal and visual form of the work being performed
- Competent by training and experience in the field, which is represented by the work in question

The early versions of the **E1527** (1993–2000) have always required that the EP conduct the site visit and the interviews. Although the EPA AAI regulation did not include this requirement, ASTM committee members drafting the 2005 version felt it was important to require some minimum standard for those conducting the site visit and the interviews. The ASTM Standard dropped the requirement that the EP performs the site visit and interviews, but it required qualifications similar to those previously used by the prior **E1527** versions to define an EP as shown in **Fig. 4.7**. This is a situation in which the ASTM Standard is more stringent, or at least more specific, than AAI.

E1527 does not include a specific section listing the EPs' responsibilities. The responsibilities are specified throughout the various sections of **E1527**. The EP conducts the Phase I ESA in accordance with the contractual agreement with the user. If the contractual agreement specifies **E1527**, then it is the EP's responsibility to perform the Phase I ESA in accordance with the standard. As such, the Phase I ESA has four main components:

1. Records review
2. Site reconnaissance
3. Interviews
4. The report

Conceptually, as depicted in **Fig. 4.8**, the Phase I ESA process can be viewed as a jigsaw puzzle. The first three components provide many pieces of the puzzle. Using the report, the EP arranges all of the pieces in the best possible fit. The opinion section of the report provides an interpretation of the missing pieces.

E1527 does not require any collection of samples. The EP is not required to perform any sampling or engage in any activities that would generate new data. The Phase I ESA consists of physical observation using sight and smell and the interpretation of existing information.

The EP is responsible for overseeing the writing of the report as well as a review and interpretation of all of the information that forms the basis of the report. There may be numerous parties associated with various portions of the ESA, and the process may involve numerous EPs. It is important to identify the sources and EPs or other parties responsible for particular portions of the process.

The Phase I ESA process involves numerous information-gathering components. The bulk of the information comes from independent businesses, governmental agencies, and public libraries. Although the EP can rely on the information provided without independently verifying its accuracy, it is important to consider it together with other findings. If other facts indicate that some source information is obviously incorrect, then the EP has the responsibility to identify such discrepancies to the user. For example, if during the site reconnaissance the EP uncovers wrong address information associated with adjoining properties and the information indicates potential RECs with respect to the subject property, then the user should be notified.

FIG. 4.7 ASTM requirements for interviews and site reconnaissance.

> The *interviews* and *site reconnaissance* shall be performed by a person possessing sufficient training and experience necessary to conduct the *site reconnaissance* and *interviews* in accordance with this practice and having the ability to identify issues relevant to *recognized environmental conditions* in connection with the *property*. At a minimum, the *environmental professional* must be involved in planning the *site reconnaissance* and *interviews*.

FIG. 4.8 Components of Phase I process puzzle.

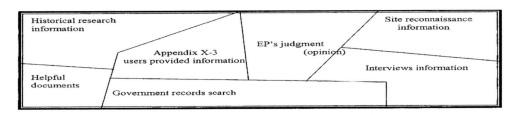

4.5 Summary

In this chapter, we reviewed some of the responsibilities the users have in providing applicable information to the EP. We reviewed the definition of the EP and considered some of the general responsibilities of the EP. We identified that the primary responsibility of the EP is to conduct the Phase I ESA. This responsibility involves thorough knowledge and appropriate application of the **E1527** Standard Practice.

References

[1] Standards and Practices for All Appropriate Inquiries, Federal Register Vol. 70, No. 210, Thursday, November 1, 2005/Rules and Regulations.

[2] ASTM Standard **E1527**, Standard Practice for Environmental Site Assessments: Phase I Environmental Site Assessment Process, *Annual Book of ASTM Standards*, ASTM International, West Conshohocken, PA, 2013.

Chapter 5 | Records Review

5.1 Introduction

One of the most important required elements of the Phase I Environmental Site Assessments (ESAs) involves obtaining and reviewing records. All of the other assessment activities, such as the interviews and the site reconnaissance, rely heavily on information uncovered in the records review portion of the Phase I ESA. Thus, the records review forms the foundation of the Phase I ESA.

The objective of the Phase I ESA is to identify recognized environmental conditions (RECs) associated with the subject property. In turn, this objective drives the purpose for the records review. Through records review, the environmental professionals (EPs) are attempting to identify any records that will help them identify any RECs in connection with the subject property.

5.2 Principles Behind E1527 Records Information

The records applicable to the requirements of **E1527** [1] must be "reasonably ascertainable." **E1527** defines this term, and we will briefly discuss the underlying principles behind it. Many records associated with the subject property fall into the reasonably ascertainable category, but EPs are only seeking records useful in identifying RECs. Combining these two principles substantially reduces the number of the records that need to be reviewed in the Phase I ESA process.

5.2.1 REASONABLY ASCERTAINABLE

Reasonably ascertainable records consist of information that is "publicly available," obtainable within reasonable time and cost, and "practically reviewable." **E1527** defines each principle further.

Publicly available information consists of any information that can be obtained by anyone who asks for it. For example, most of the government information related to environmental regulations is considered publicly available. Internal corporate documents associated with the manner in which the neighboring property processes wastes would not be considered publicly available. However, if the corporation is listed as a waste generator, then the information that is reported to the government is considered publicly available.

Information obtainable within a reasonable time consists of information that can be obtained from the source within 20 calendar days from the time of request. Many Phase I ESAs are conducted in less time than 20 days. If the necessary information cannot be obtained in the allotted time for the ESA, then the EP must determine whether the information was not reasonably ascertainable or the research was limited by time constraints established by the user. **Figure 5.1** shows the decision-making approach. The example assumes that the Phase I ESA must be completed in 10 days. If the information was ascertainable in 20 days but not in 10, then the failure to review those records would be listed as a limitation to the Phase I ESA. This limitation may also be a significant "data gap" if the lack of information is determined to affect the EP's decision regarding a REC.

E1527 defines reasonable cost as the sources' nominal cost of retrieving and duplicating. Some records may be available for the property, but the cost of retrieval may be high. In cases in which the retrieval costs are excessive, the information is not considered reasonably ascertainable. Discussion with the user is appropriate and may be an important part of the contractual agreement and subsequent communications.

The standard also defines "practically reviewable." This definition addresses the form in which the information is available for review. The information must be relevant to the environmental aspects of the property and should not require any extraordinary analysis of the data. For example, if the information is only identified by zip code and the subject property is located in an urban area, then it would require too much effort to crosscheck all of the listings against the address information. This would be an example of information that is considered not practically reviewable.

5.2.2 OBTAINING THE RECORDS

The EPs must make a reasonable effort to obtain the records. Even if the records are not reasonably ascertainable, the EPs must then attempt to obtain the required information through the other activities associated with the assessment, such as interviews with the owners, occupants, and users.

EPs can engage a non-EP in the activity of gathering the records or order the record search from a commercial service. Because it is their business, most commercial service companies that specialize in providing environmental information to consultants are far better equipped to generate the information than the EPs. With fast

FIG. 5.1 Decision process for reasonably ascertainable information.

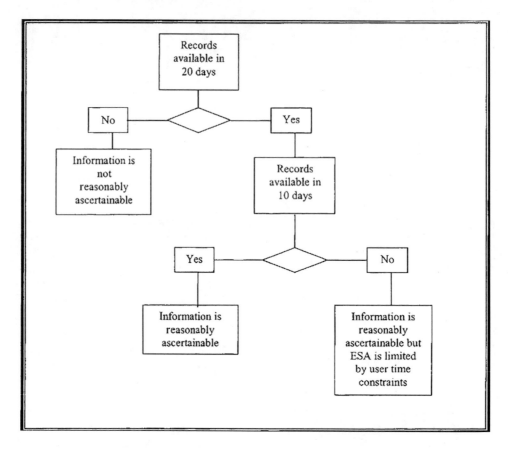

turnaround times, user-friendly format, competitive price, and timely updating in conformance with the requirements of E1527, the reports streamline some of the Phase I ESA records review process.

The EPs can rely on the various information sources obtained without being obligated to check it for accuracy and completeness. Only in cases in which other activities of the Phase I ESA uncover mistakes or insufficiencies are the EPs obligated to make an effort to consider the effect of those discrepancies in light of the environmental conditions of the subject property.

E1527 requires the EP to document each source that was used, even if it did not reveal any findings. The sources must be sufficiently documented, including name, date of request, and the date the information was last updated by the original source. The EPs generally satisfy this requirement by including a complete copy of the record in the appendices. If the EPs only provide a reproduced portion of an aerial photograph or a map in the report, then they must take additional steps to separately document relevant source information, such as name and date.

5.3 Mandatory and Optional State and Federal Record Information

E1527 requires several standard environmental records sources be included in the research and specifies minimum search distances from the property boundary. These requirements are shown in **Fig. 5.2**. The figure also includes the required minimum search distances. Minimum search distances relate to the nature of the database. The search radii for databases of contaminated sites are greater than for databases for sites that have only generated hazardous waste or have emission permits. An understanding of what each database represents and is tracking is an important element of assessing database search results.

Minimum search distances are measured from the property boundary. Commercial sources often provide radius searches with the property center forming the central point of the searches. **Figure 5.3** depicts a typical map generated through a radius search. The EPs should pay particular attention to the minimum search distance requirement. Larger or irregularly shaped properties may require the EP to request an extended radius search to meet the **E1527** minimum search distance requirement.

As previously discussed, EPs typically order a record search from a commercial service. Commercial services provide information that is contained in regulatory agency databases. However, database information typically presents excerpts of agency file information. Because of a growing trend of EPs relying entirely on the commercial database-provider information when forming opinions about RECs, the ASTM task group developed minimum guidance for reviewing the pertinent regulatory files associated with certain listings.

Agency file review requests have routinely been made for the subject property as an initial step in the assessment process.

FIG. 5.2 Mandatory federal and state sources.

Standard Environmental Record Sources (where available)	Approximate Minimum Search Distance miles (kilometres)
Federal NPL site list	1.0 (1.6)
Federal Delisted NPL site list	0.5 (0.8)
Federal CERCLIS list	0.5 (0.8)
Federal CERCLIS NFRAP site list	0.5 (0.8)
Federal RCRA CORRACTS facilities list	1.0 (1.6)
Federal RCRA non-CORRACTS TSD facilities list	0.5 (0.8)
Federal RCRA generators list	property and adjoining properties property only
Federal institutional control/engineering control registries	
Federal ERNS list	property only
State and tribal lists of hazardous waste sites identified for investigation or remediation:	
State- and tribal-equivalent NPL	1.0 (1.6)
State- and tribal-equivalent CERCLIS	0.5 (0.8)
State and tribal landfill and/or solid waste disposal site lists	
State and tribal leaking storage tank lists	0.5 (0.8)
State and tribal registered storage tank lists	property and adjoining properties property only
State and tribal institutional contral/engineering control registries	
State and tribal voluntary cleanup sites	0.5 (0.8)
State and tribal Brownfields sites	0.5 (0.8)

Source: Adapted from **E1527-13**.

The new wrinkle with the 2013 standard is that adjoining property addresses now must also be considered if listed on one or more of the mandatory record sources. With the end product in mind (Phase I ESA report and U.S. Environmental Protection Agency [EPA] All Appropriate Inquiries Final Rule [AAI] compliance), EPs need to be cognizant of the actual time it takes to complete agency file reviews. Many times these are written requests to government agencies that do not move at the speed of business, and some forethought and planning are prudent to make sure you get the information you need within the time frame you have been granted to complete your Phase I ESA report. Other times they are easily obtained through online records or are available over the counter. Effort should be made to understand all known addresses for the subject property to complete a full search. If historical addresses are encountered during the Phase I ESA, then those also need to be considered by the EP.

In the case of agency file reviews for adjoining properties, the EP must get the adjoining property addresses, review if any are within the database report for listings that warrant supplemental information gathering, and then submit requests or retrieve the agency file information—all within the time allotted for the completion of the Phase I ESA. Agency file reviews are a critical path item that should be carefully managed. Many EPs are aware that some agencies charge a fee to retrieve records and should account for this eventuality in relation to extra requests for adjoining properties beyond the routine gathering of the subject property address.

Figure 5.4 presents the requirements outlined in the **E1527-13** regarding agency file reviews. Note that the standard states that the file reviews "should" be conducted, but if a review is not conducted, the EP "must" provide justification in the report for not conducting the review. For example, the EP may supplement the mandatory information by checking several other sources. Additional sources are usually included in the Phase I ESA if the EP has found the source to be useful in the past or if it is a part of local customary practice. **Figure 5.5** lists several examples of the types of local records and local sources of information.

Commercial sources of environmental information can also provide information that may be useful for the EP in researching the property. The services usually provide information about the information sources searched, including agency release dates. The types of available information will vary from state to state, and some of the information may be only available on the local level.

Many commercial search services offer additional nonmandatory databases as part of their standard Phase I package. Some also offer nondatabase information, such as maps, physical setting information, or historical data. These services are normally provided at an extra cost.

If the EP elects to receive additional nonmandatory databases, then they need to be reviewed. Adding a significant number of additional databases can greatly increase the review time required

FIG. 5.3 Example of records research radius map.

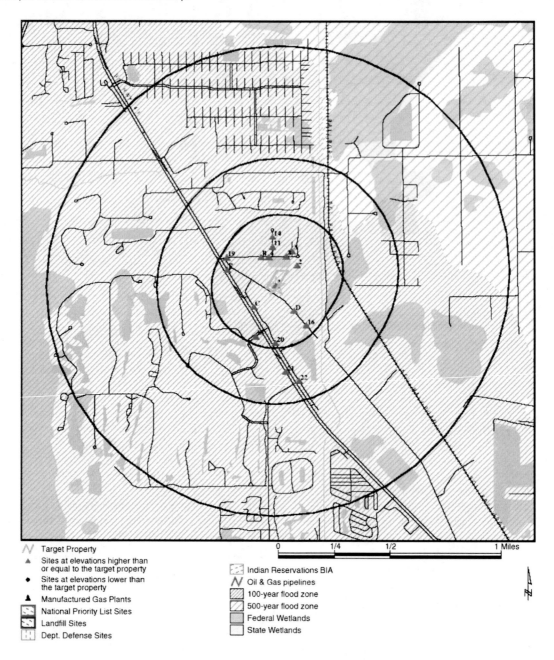

FIG. 5.4 Guidance regarding agency file reviews.

> If the *property* or any of the *adjoining properties* is identified on one or more of the standard environmental record sources, pertinent regulatory files and/or records associated with the listing should be reviewed. The purpose of the regulatory file review is to obtain sufficient information to assist the *environmental professional* in determining if a *recognized environmental condition*, *historical recognized environmental condition*, *controlled recognized environmental condition*, or a *de minimis condition* exists at the *property* in connection with the listing. If, in the *environmental professional's* opinion, such a review is not warranted, the *environmental professional* must explain within the report the justification for not conducting the regulatory file review.

Source: Adapted from **E1527-13**.

FIG. 5.5 Local record types and sources.

> Types of records
> - Local Brownfields lists
> - Local lists of landfill/solid waste disposal sites
> - Local lists of hazardous waste/contaminated sites
> - Local lists of registered storage tanks
> - Local land records (for AULs)
> - Records of Emergency Release Reports (42 U.S.C. 11004)
> - Records of contaminated public wells
>
> Sources
> - Department of health/environmental division
> - Fire department
> - Planning department
> - Building permit/inspection department
> - Local/regional pollution control agency
> - Local/regional water quality agency
> - Local electric utility companies (for records relating to polychlorinated biphenyls)

Source: Adapted from **E1527**.

to evaluate the information. A good rule of thumb is that if the EP is not going to look at it, then they should not ask for it.

One principle to keep in mind is that the purpose of the records review is to obtain information related to RECs, historical recognized environmental conditions (HRECs), and controlled recognized environmental conditions (CRECs). Information that provides new insight to the property is useful. Additional information that simply confirms what is already known is of limited value. For example, if the EP has three sources that indicate that there was an underground storage tank (UST) on the site, then an additional source that confirms that there was a UST on site is of little value.

Sometimes the nonmandatory information includes a nonscope item. EPs should be cautious about obtaining and including data that address nonscope items such as radon or wetlands in their Phase I report unless it is specifically included in the scope of work negotiated with the client. Including such additional information implies that it is the consultant's standard practice and exposes the consultant to liability for the nonscope information.

5.4 Mandatory and Optional Physical Setting Records Information

It is important to recognize that a REC only applies to a release that affects the subject property. A release that has occurred a block away but is not expected to affect the subject property is not a REC. Thus, the purpose of reviewing physical setting sources is to gain an understanding about the geologic, hydrogeologic, hydrologic, or topographic characteristics of a site. This information is important in evaluating the potential for contaminants to migrate to or from the subject property.

FIG. 5.6 **E1527-13** definition of migration.

> *Migrate/migration:* For the purposes of this practice, "migrate" and "migration" refer to the movement of *hazardous substances* or *petroleum products* in any form, including, for example, solid and liquid at the surface or subsurface, and vapor in the subsurface.

The evaluation of migration pathways has always been important in developing opinions about RECs, but it was not until the 2013 version of **E1527** that a definition of "migration" was included in the standard (see **Fig. 5.6**).

E1527 requires the EP to review a current topographic map to satisfy the physical setting source requirement. Additional physical setting sources can be provided to further supplement the information and are required when migration of contamination is likely or additional sources are commonly used in local practice.

A list of mandatory and optional physical setting sources is presented in **Fig. 5.7**. Sometimes the development plans by a construction/engineering firm are available to the EP. They can be an additional useful site-specific physical setting source.

5.5 Historical Research

The purpose of the historical research is to consider the possible releases that might have occurred at a property in light of the historical uses and, in concert with other relevant information gathered as part of the Phase I process, to use this information to assist in identifying RECs.

E1527 requires the EPs to conduct research into the historical uses of the property to the first developed use or 1940, whichever is the earlier. For example, if the property was first developed for agriculture in 1850, the research would begin with 1850. If the property was first developed for residential property in 1960, then the research would have to go back to 1940. The research is limited to issues that could potentially be associated with RECs. The EPs only have to identify general types of use unless the research uncovers industrial uses that may have resulted in contamination of the property. Industrial uses dictate further research into specific industrial classification and potential contaminants associated with it.

Figure 5.8 lists the eight standard historical sources outlined in **E1527**. Any one or any combination of the sources in the table may be used to satisfy the requirement of the standard. For

FIG. 5.7 Physical setting sources.

Physical Setting Sources	
• **Mandatory**	Current 7.5 minute topographic map
• **Optional**	Groundwater maps Bedrock geology maps Surficial geology maps Soil maps Other credible sources

Source: Adapted from **E1527**.

FIG. 5.8 Standard historical sources.

Standard Historical Sources

- Aerial photographs
- Fire insurance maps
- Property tax files
- Recorded land title records (cannot be the sole source)
- Historical U.S. Geological Survey topographic maps
- Local street directories
- Building department records
- Zoning/land use records
- Other historical sources (e.g., miscellaneous maps, newspaper archives, Internet sites, community organizations, local libraries, historical societies, interviews)

Source: Adapted from **E1527**.

example, if aerial photographs are available for the history of the property, then reviewing this source alone will technically suffice. The only stated exception to this rule is associated with recorded land title records. Because the title research usually provides only ownership information and minimal information with respect to the actual use of the property, one additional source must be used in conjunction with this source. However, practically speaking, a combination of historical sources is typically required to satisfy the objectives of the historical research. For example, in urban areas, city directories provide information for specific address locations; however, address numbers frequently change. Used in combination with fire insurance maps, a more complete and accurate historical record can be established. In rural areas, aerial photographs can depict lakes, mountains, and vast tracks of apparently undeveloped land; however, historical topographic maps identify settling ponds, old mines, and other features not necessarily apparent on high-flight photographs.

The historical research is only required for the subject property; however, the uses of the surrounding area should be provided in the report to the extent they are identified in the course of the Phase I ESA.

E1527 does not require EPs to research the uses in intervals of less than 5 years. If the information is not reasonably ascertainable, then gaps in the research can be explained as "data failure." If the research shows that no significant change in the use of the property took place at longer than the 5-year intervals, then the standard does not require any further research. For example, if an aerial photograph shows an apartment building on the site in 1954, and the same building is still on the site when the Phase I ESA is being prepared, no intervening intervals need to be researched.

When it is not possible to determine the land use history, any gaps or intervals longer than the 5-year interval should be explained in the report. For example, it is not uncommon to have data failure because sufficient records do not exist for land uses before the early 1900s.

Many of the historical sources outlined in **E1527** may not provide the EP with useful information with respect to the environmental conditions of the subject property. Such sources can be specifically excluded from the research on the basis of the EP's past experience. In some cases the EP may find that the information is not available or is incomplete from a particular source. For example, the EP may have found through prior research that property tax assessor records in certain jurisdictions do not predate current site uses or that fire insurance maps are not available for rural areas or a particular town. This lack of or inability to obtain information from a particular source is data failure. The concept of data failure has traditionally been a part of the ASTM historical research.

The EPA AAI regulation introduced a separate concept—the data gap. The EP should not confuse data gaps with data failure. Data failure was introduced to deal with the fact that it is often difficult or impossible to identify all of the property uses back to the original development. The concept is used to bring closure by noting the problem but satisfying the technical requirements of the standard. Data failure is a type (thus a subset) of a data gap.

5.6 Summary

In this chapter, we reviewed the mandatory and nonmandatory records research elements associated with Phase I ESAs. We discussed the federal and state records research, physical setting research, and historical research of **E1527**. We looked at some of the limitations and options the EPs have in meeting the requirements of the standard.

References

[1] ASTM Standard **E1527**, Standard Practice for Environmental Site Assessments: Phase I Environmental Site Assessment Process, *Annual Book of ASTM Standards*, ASTM International, West Conshohocken, PA, 2013.

Chapter 6 | Site Reconnaissance

6.1 Guiding Principles of Site Reconnaissance

Site reconnaissance is required by the **E1527** Standard. **E1527** [1] uses two terms associated with site inspection. They are "site visit" and "site reconnaissance." Site visit can be considered as a subset of site reconnaissance. Site visit generally refers to inspection activities of the subject property. Site reconnaissance includes those activities but expands the event to the analysis of site visit information in concert with other information in identifying recognized environmental conditions (RECs).

The environmental professional (EP) must visually and physically observe the property and any structures on the property. This entails more than just a drive-by inspection of the property and snapping a few photos out of the car window. Photographs with the rear view mirror showing in the corner (see **Fig. 6.1**) are difficult to explain to an observant user and can be perceived as an indicator of inadequate effort on the part of the EP. Should the EP have the misfortune of being involved in a legal dispute with respect to the assessment of the property, any details that can be interpreted as a lack of professionalism are readily picked up by the opposing attorney in an attempt to discredit the quality of the EP's work product.

E1527 recognizes the limiting conditions that the EP may encounter during site reconnaissance and provides guidance with interpretation of the term "visually and physically observed" in the standard. During a visit, portions of the property may be obstructed by bodies of water, rock formations, ravines, adjacent buildings, snow, vegetation, and other obstacles that may impede the EP's ability to detect RECs or other significant features of the subject property. The EP should note these physical limiting conditions in the report.

Local knowledge is important. From a business and logistical perspective, it is preferable for the EP to establish a geographical area that they are going to service and become familiar with its environment, local regulations, and local practices. The EP responsible for the report is sometimes not familiar with the general area or is out of state. This can be a constraint to the site visit because of the limited local knowledge of the EP. In some cases, this can lead to inaccurate assumptions about the property. For example, consider a subject property along a river. The out-of-state EP visits the property and notices the adjoining industrial property to the south discharging effluent into the river. A map indicates the industrial site as downstream of the subject site, and during the site visit the EP's observation of the flow of the river confirms the map information. In his report, the EP dismisses the potential for the effluent to migrate upstream onto the subject property. Several weeks after the purchase of the property, the user finds that the banks of the subject property are contaminated by the discharge from the "downstream effluent." That portion of the river is affected by tidal flow, and 2 hours after the EP left the property the flow of the river reversed.

Experience and training of the EP is also very important. The signs of historical uses or activities may often be overlooked simply because the EP does not know what they are looking at. Some of the knowledge can be gained from experience, self-education, attending training sessions, and mentoring. A good rule is to be observant, and if the EP sees something that they do not understand, then they should inquire further. If you do not know, then ask.

In some cases, in an attempt to minimize travel and local knowledge research costs, the out-of-town EPs subcontract the site reconnaissance to a local EP. These facts should be disclosed to the user, and all EPs' credentials should be provided. Subcontracting portions of the Phase I Environmental Site Assessment (ESA) tasks can significantly complicate liability issues. Consulting firms should review their subcontracting policies to ensure sufficient protection from liability.

E1527 does not require the EPs to conduct the records review, interviews, and the site reconnaissance in any specific order. Ideally, when the EPs have most of the information from the records review before physically investigating the site, they are better prepared for the site investigation. In the real world, time limitations and other transaction-imposed constraints often require some of the assessment activities to take place simultaneously or out of the preferred sequence. This may necessitate a second visit to the site. The standard does not require multiple site visits, but from a practical point of view it is a good strategy to conduct the communications and site reconnaissance activities in a friendly manner to facilitate a return to the site should it become necessary.

In some cases, the property may have been previously assessed and the EP was fortunate enough to receive the prior assessment before conducting the site reconnaissance. Such information can be extremely useful, but the EPs cannot rely on the information in its entirety and must conduct another site visit. Over-reliance on prior information can focus the attention of the EPs in areas

FIG. 6.1 Drive-by is not sufficient to meet the requirements of **E1527**.

Source: Provided by J. Kilgore.

previously identified as potential problems and can result in some RECs being missed. Prior information should not deter the EPs from conducting a full and complete investigation. Most EPs develop a systematic approach to site visits and consistently use that approach. The approach can be as simple as using the transaction screen questionnaire from **E1528** [2] to ensure that the areas covered by the standard are addressed. Another method is to review Section 9 of **E1527** and generate a list of items to note during the site visit. **Table 6.1** is an example of the list.

The list in **Table 6.1** is not all-inclusive. Local knowledge and the particular expertise of the EP may generate additional categories depending on the characteristics of the subject property and local practice. All of the items in the list in **Table 6.1** are included and specifically mentioned in Section 9 of **E1527**. At a minimum, these observations should be a part of any Phase I ESA. The EPs do not necessarily have to write a paragraph of notes on each of the 28 items, but through various documentation techniques, such as systematic photography, the EPs should address each of the points listed. The EPs should select a site reconnaissance methodology that does not leave any of the points listed in **E1527** out of the documentation of the report.

The **E1527** site reconnaissance includes three compulsory components: general setting, exterior observations, and interior observations. It is helpful to arrange these three components in a logical order, starting with the observations of the area and gradually reducing the focus to specific items associated with and potentially affecting specific portions of the subject property. The information gathered during the site reconnaissance must be considered together with all of the other information obtained during the ESA (i.e., historical research, interviews, and specialized

TABLE 6.1 Observable conditions during site reconnaissance

Property identification:

1. Methodology for walk over:
2. Limiting conditions:
3. General setting/neighboring properties:
North:
East:
South:
West:
4. Current uses subject property:
5. Current uses neighboring property:
6. Current uses surrounding area:
7. Observed indications of past uses subject property:
8. Observed indications of past uses neighboring property:
9. Observed indications of past uses surrounding area:
10. Geologic, hydrogeologic, hydrologic, and topographic conditions:
11. Structures and improvements on subject property:
12. Roads:
13. Water supply:
14. Sewage disposal system:
15. Storage tanks:
16. Chemical storage:
17. Drums/containers:
18. Solid waste disposal:
19. Pools of liquid:
20. Ponds, pits, lagoons:
21. Odors:
22. Polychlorinated biphenyls electrical/hydraulic:
23. Heating/cooling:
24. Drains/sumps:
25. Leaks, stains, corrosion:
26. Stressed vegetation:
27. Wastewater/cooling water:
28. Wells:

knowledge). Often, it is only after the parts are considered as a whole that information about a REC will surface or be dismissed.

6.2 General Site Setting

The EPs must understand that the Phase I ESA extends beyond the boundary of the subject property. The uses of the surrounding area may potentially affect the environmental condition of the subject property. Contamination from surrounding properties can migrate onto the subject property by way of groundwater migration, vapor migration, or aerial deposition.

Under "general site setting," **E1527** requires the EPs to review the current and past uses of adjoining properties. Let us take a closer look at the requirements of the standard in this respect. According to **E1527**, "adjoining property" is any real property or properties, the border of which is contiguous or partially contiguous with that of the property, or would be contiguous or partially contiguous with that of the property but for a street, road, or other public thoroughfare separating them. This concept is illustrated in **Fig. 6.2**. For example, in the report, the EP writes that the adjoining property to the west is a six-lane highway without providing any information about the property on the other side of the highway. This would constitute failure to meet the requirements of the **E1527** standard.

The EP is required to identify past uses of the adjoining properties. **E1527** Section 9.4.1.4 explains that identification of past uses of the adjoining properties during site reconnaissance consists of

- Physically observed items during the site visit,
- Information gained from interviews, and
- Records research.

FIG. 6.2　**E1527** concept of adjoining property.

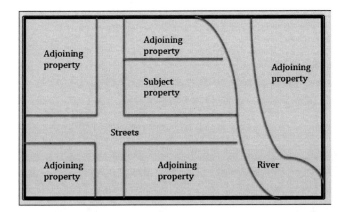

The physically observed items can include old signs on neighboring buildings (e.g., ABC Paint Factory, AAA Transmission Repair, Lee's Dry-Cleaning), neighboring building characteristics (multiple automotive repair bays, power plant, sewage treatment plant), or waste materials dumped on the adjoining properties (automotive batteries, tires, drums, piles of dirt, tank farms). **Figures 6.3–6.21** show several examples of conditions that the EP may encounter.

The interviews can also provide information about surrounding properties. **E1527** intends that the EP should ask the questions about the surrounding area, not just the subject property, and should confirm the validity of the answers through physical observation and/or records review. **Figure 6.8** is an example in which interviews and physical observations can reveal significant detail about a REC. The tenant of a commercial strip building could not

FIG. 6.3　Although these buildings are currently used as a church and a restaurant, on-site observations clearly indicate different uses in the past.

Source: Provided by J. Kilgore.

recall what he used to clean motorcycle parts. Closer inspection of a stack of 5-gal containers revealed that a dry-cleaning solvent was used, and several spills were observed outside of the shop door.

A frequently asked question is, "How much effort is required of the EP when identifying the uses of the adjoining properties, especially with respect to the historical use of those properties?" Only current uses that are visually or physically observed or identified through interviews or records research need to be identified in the site visit section of the report. With respect to the historical uses of adjoining properties, Records Review Section 8.3.3 of **E1527** provides the answer to that question. The EP has to identify historical information about the surrounding area to the extent that it is revealed in the course of historical research of the subject property. **E1527** also allows the EP to use discretion in limiting the extent of the effort. The EP is not required to independently research the uses of the adjoining properties to 1940 or the earliest development. That requirement only pertains to the subject property.

The underlying principle of the standard is to consider and use the information gathered in the site reconnaissance together with the information gathered in other portions of the process, such as the interviews and/or historical records research.

Another frequently asked question in this area deals with the historical subdivision of properties. **E1527** considers only the current (at the time of the Phase I ESA) boundary. If the property was a portion of a larger parcel in the past, then the standard does not require the EPs to continue to expand the research boundary with every subdivision event.

FIG. 6.4   The EP must document the current and historical source of heat. Although this building is currently connected to natural gas, the old heating oil tank was still present at the back of the building.

Source: Provided by J. Kilgore.

FIG. 6.5A   Nuisance dumping typically affects the property because of the cost associated with removal, but it typically is not a REC.

Source: Provided by J. Kilgore.

FIG. 6.5B   Other dumping is more concerning.

Source: Provided by J. Kilgore.

FIG. 6.6 The EP must document any limiting conditions that could obscure observations such as snow or thick vegetation.

Source: Provided by J. Kilgore.

6.3 Exterior and Interior Observations

The approach presented in the previous section on general site setting is also applicable to this section of the site reconnaissance. The **E1527** Standard further subdivides the exterior and interior observations into the following three categories: general observations, interior observations, and exterior observations. The intent is to identify historical and current uses likely to involve the use, treatment storage, disposal, or generation of hazardous substances. These must be identified by the EP in the report.

FIG. 6.7 The contents of tanks, drums, and containers must be identified.

Source: Provided by J. Kilgore.

6.3.1 GENERAL OBSERVATIONS

The information about the uses of the property is considerably more helpful to the user when it specifically identifies the use. For example, instead of retail use, the EP should specify that it is a bakery or automotive battery retail store. In a small strip plaza, or properties with multiple tenants, it can be helpful to take a photograph of the mall sign listing the tenants as shown in **Fig. 6.9**. Signs are often helpful in providing information about current or past uses of the property. The business activities may have ceased, but the sign remains.

These and other evidence of current or historical uses noted during the site reconnaissance (or records research or interviews or both) must be provided to the user in the report. Section 9.4.2 of **E1527** specifically requires the EP to note

- Hazardous substances and petroleum products in connection with identified uses
- Hazardous substances and petroleum products not in connection with identified uses
- Unidentified substance containers

FIG. 6.8   Containers of a parts-cleaning liquid turned out to be a "dry-cleaning solvent" that had been spilled or dumped outside of the shop bay.

Source: Provided by J. Kilgore.

FIG. 6.9   Sign listing tenants.

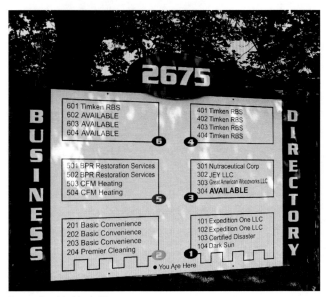

Source: Provided by J. Kilgore.

FIG. 6.10   Exterior observations: aboveground storage tanks with evidence of release.

Source: Provided by J. Kilgore.

- Drums
- Pools of liquid
- Odors
- Storage tanks
- Polychlorinated biphenyls (PCBs; light ballasts are excluded)

The details of Section 9.4.2 are not repeated here. EPs need to carefully review this section and the rest of the standard to develop a thorough understanding of their reporting responsibilities with respect to **E1527**.

6.3.2   EXTERIOR OBSERVATIONS

The observations of the exterior of the subject property must be documented in the report. Exterior observations that must be described in the report are listed in Section 9.4.4 of **E1527** and include

- Pits, ponds, lagoons
- Stained soil or pavement
- Stressed vegetation

FIG. 6.11 Surface staining leading to a storm drain collection grate.

Source: Provided by J. Kilgore.

FIG. 6.12 Interior observations of a staining that may or may not constitute a REC.

Source: Provided by J. Kilgore.

FIG. 6.13 Multiple underground installations.

Source: Provided by J. Kilgore.

- Solid waste
- Wastewater
- Wells
- Septic systems

EPs need to carefully review this section of the standard to develop a thorough understanding of their reporting responsibilities with respect to exterior observations.

EPs' experience and local knowledge of historical uses of the general area can be invaluable in helping to identify potential problem areas. For example, familiarity with local industries may provide clues about wastewater systems. Local knowledge of plants' biology can provide clues about stressed vegetation that may be caused by natural cycles and not necessarily by contamination of the subject property.

The EPs must also pay particular attention to any unusual observations during the site visit. Piles of fill dirt, berms, dead end roads, gullies, and other unusual topographic features should be noted. These may be associated with historical uses of the property, and the EPs should attempt to gain information about the reasons for their presence from the interviews. **Figure 6.11** depicts a photograph of surface staining leading to a storm drain collection system. Tracing the source of this staining back to its source, the EP found a parts-cleaning sink inside of the paint room that drained

FIG. 6.14 Some tanks are more difficult to identify. This fill port and vent line are associated with a heating oil tank.

Source: Provided by J. Kilgore.

FIG. 6.15 Evidence of underground lift storage tank repair. Lift storage tanks typically do not require registration because of their small size. Some of the older tanks used PCB-containing oils. Many of these installations were prone to leaks.

Source: Provided by J. Kilgore.

to a small tank with a submersible pump that discharged the wastewater to this storm drain.

6.3.3 INTERIOR OBSERVATIONS

Interior observations are to be made with the intent to identify releases to the environment. An observation of staining on an interior concrete floor is to be noted in the report, but unless there is a means for this spill to make its way into the environment (through drains, sumps, flooring, cracks, etc.), it would not rise to the level of a REC.

Interior observations are limited to accessible common areas—those expected to be used by the occupants or the public. It is not

FIG. 6.16 Evidence of former drums in secondary containment with a drain that discharged to a nearby storm drain retention basin.

Source: Provided by J. Kilgore.

necessary to look under floors, above ceilings, or climb onto roofs. In addition to the observations listed above, Section 9.4.3 (Interior Observations) of **E1527** specifically requires the EP to identify

- Heating/cooling
- Stains or corrosion
- Drains or sumps

Again, the EPs need to carefully review this section and other sections of the standard to develop a thorough understanding of their reporting responsibilities with respect to **E1527**. The observations should be described and documented in sufficient detail to provide the user with a thorough understanding of observed conditions and how they may potentially affect the environmental conditions of the subject property. Whenever possible, the EPs need to evaluate the exterior and interior observations together. For example, an interior sump pump observation should be evaluated with any exterior noted discharges and impacts such as stressed vegetation.

With respect to site reconnaissance, **E1527** requires the EP to document the methodology of the walkover. Documentation of the methodology enables the user to understand exactly what was done, and it provides future reference for the EP. It is possible to miss a REC because of the methodology used in the walkover, especially with large properties. Describing the methodology with sufficient detail provides information to the user about a potential for missing a REC.

In a recent Phase I ESA, the EP described his methodology for a walkover as "via the transverse method, whereby the property is walked in a random pattern." The report included six photographs

FIG. 6.17 Abandoned monitoring well in a parking lot.

Source: Provided by J. Kilgore.

FIG. 6.18 A GPS can be used to document precisely where the EP walked during the site reconnaissance and downloaded as an overlay on an aerial photograph. This is particularly useful for large tracts of land with very limited access.

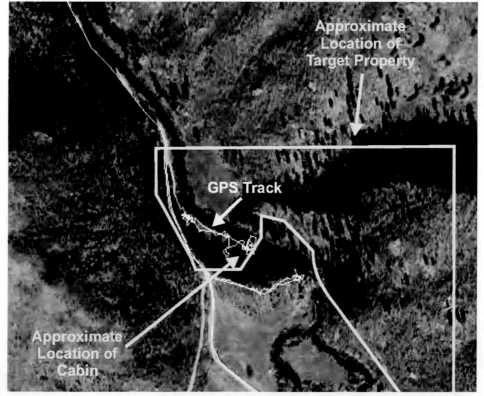

Source: Provided by J. Kilgore.

FIG. 6.19 Evidence of mine tailings raise concerns about elevated metals that are likely to be present where the mine tailings are visible and where aerial deposition and surface runoff would accumulate.

Source: Provided by J. Kilgore.

FIG. 6.20 Elevated metals are often associated with sandblast grit.

Source: Provided by J. Kilgore.

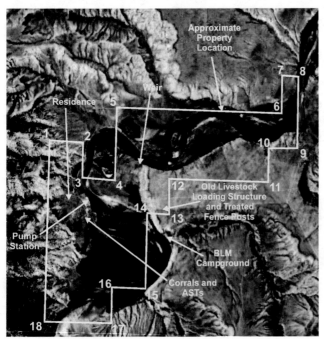

FIG. 6.21 An aerial photograph is used in combination with numbered photographs presented in the Phase I ESA report to document where a particular photograph was taken and the direction of the view.

Source: Provided by J. Kilgore.

FIG. 6.22 Documenting transformer with no PCBs.

Source: Provided by J. Kilgore.

of the 5-acre property. This methodology description did not sufficiently describe to the user which parts of the property the EP actually inspected. When asked if the EP walked over certain portions of the property, the EP could not remember. Documentation should always be adequate to provide a method for the EPs to retrace their actions. With larger properties, it is helpful to identify the method in greater detail. For example, "using a grid pattern with 100-yd intervals" or "north and south transverses on 100-ft centers," and identifying the starting location tells the user where the EP walked during site reconnaissance.

This method can be used in combination with numbered photographs taken at a specific point during the walkover. A description of the direction of the view in each photograph further documents the methodology used. For smaller properties, a descriptive photographic method can be sufficient.

Except for field notes, **E1527** does not specify or recommend a specific method of documentation. Numerous methodologies have been used by consultants, including video, tape recorders, and standard and digital photography. It is up to the EP to select a method that meets the needs of a particular ESA. Different ESAs may dictate different forms of documentation. Note that photographs are not required as part of the **E1527**.

Photography has by far been the most widely used method of documentation in Phase I ESAs. The old cliché "a picture is worth a thousand words" is appropriate here. The EP's time is valuable. It takes only a fraction of a second to take a photograph of a property condition that could literally take hours to describe in field notes.

Photographs also help with the interpretation of the language. For example, "stained pavement" is required to be described in the report. A photograph of the stain enables the user to use judgment without having to rely solely on the EP's interpretation. Photography is a valid and useful tool, but like any other tool it is prone to equipment failure and/or operator error. It is appropriate to consider several helpful hints on ESA photography.

Because modern cameras operate on batteries, make sure that your field kit includes extra batteries or a car charger. Electronics sometimes have a tendency to act up because of extremes of weather (heat, humidity, cold). Some EPs carry small disposable cameras just in case the "hi-tech" one fails.

How many pictures should the EPs take? The EPs should take a sufficient number of photographs to document the findings and the methods used. The documentation should enable the EP to retrace individual steps in the Phase I ESA process. Six photographs, combined with a description of "random transverse walkover" do not represent sufficient documentation of a 5-acre property. An extra photograph is considerably less expensive than revisiting the site. However, it is important to be thoughtful about which photographs should be included in a report. It is unnecessary and inefficient to include numerous photographs that are unrelated to discussions in the text of the report.

Having a sufficient number of photographs of the site and the surrounding area can also be extremely helpful when combined with aerial photograph research. The camera can sometimes capture an important finding that may have escaped the eye during site reconnaissance. Thorough review of photographs can enable the EP to recognize missed information before the completion of the Phase I ESA and enables its inclusion in the final report.

Site reconnaissance requires interior and exterior observations. In some cases, interior spaces are large with insufficient light or no light at all (e.g., abandoned warehouse). Flashes of many modern cameras, smartphones, or other electronic devices may not be capable of sufficiently illuminating large areas, resulting in dark backgrounds in the photographs. Consider carrying a more powerful external flash.

Technical photography differs from creative photography. In creative photography, the artist's goal is to evoke a feeling or an emotion. The point of Phase I ESA photography is to document a condition. To that end, the photographs should include points of reference and objects that enable the user to judge size or scale. **Figure 6.23** shows a photograph of a stain without a scale. By including an object of known size in the photograph, the viewer can judge the size of the stain. Another method would be to include a background that provides the scale, such as a 55-gal drum. Without the size reference, it is difficult to judge the size from the photograph.

EPs often use the small "Date" inserts printed in the corner of the photographs. This allows the user to see when the photos were taken. If you are using this tool, then make sure that you check the date before you start shooting. It can be embarrassing when the date of site reconnaissance in your report does not match the date on your photographs.

The photographs taken during the site reconnaissance are part of the EP's file. Occasionally, the real estate transaction falls apart and results in litigation. Overnight, the EP's files, including notes, photographs, and negatives, become evidence. File retention policies and issues such as who owns and maintains the original photographs need to be considered by consulting firms involved in ESAs. Legal arguments can be made for retaining only the pictures used in the report or for retaining all site photos. It is important that some consistent policy be adopted.

The use of digital photography is nearly universal. The comparative ease by which the image can be changed and manipulated has created several problems in evidentiary admissibility of this type of photography. Consultants need to be cognizant of the limitations.

FIG. 6.23 Stain without a scale reference.

Source: Provided by J. Kilgore.

6.4 Summary

In this chapter, we reviewed several practical implications of **E1527** Section 9 on site reconnaissance. We pointed out EPs' reporting responsibilities associated with general site setting and exterior and interior observations. We addressed some of the documentation issues and discussed the role of photography as a documentation tool during Phase I ESAs.

References

[1] ASTM Standard **E1527**, Standard Practice for Environmental Site Assessments: Phase I Environmental Site Assessment Process, *Annual Book of ASTM Standards,* ASTM International, West Conshohocken, PA, 2013.

[2] ASTM Standard **E1528**, Standard Practice for Limited Environmental Due Diligence: Transaction Screen Process, *Annual Book of ASTM Standards,* ASTM International, West Conshohocken, PA, 2014.

Chapter 7 | Safety and Health

7.1 Safety and Health Issues Associated with Site Investigations

The ASTM Phase I and Phase II Environmental Site Assessment (ESA) standards [1,2] do not provide any direction or guidance to the environmental professionals (EPs) with respect to health and safety issues. This fact should not be construed to imply that safety is not important or that it does not apply to the assessment activities. With Phase I ESA investigations, the issues can be very basic and driven by common sense. For example, the EPs are not required by the standard to enter confined spaces, and common sense tells the EPs not to engage in sniffing and taste testing of the contents of a 55-gal industrial drum. However, there have been instances in which EPs' safety was affected by factors not necessarily related to recognized environmental conditions (RECs). For example, an EP walked into an abandoned warehouse when the door stuck shut, and the EP was rescued several hours later only because her home office knew where to look for her. There are market and budgeting pressures that often drive the EP to minimize the amount of personnel to perform the tasks. Site reconasaince is often performed by a single EP. This can lead to safety issues, and some companies have found that using a team of two people can improve safety, quality control, and efficiency.

In addition to RECs, guard dogs, snakes, alligators, unfriendly neighbors, migrants, and occupants have all been encountered by the unsuspecting EPs performing ESAs. **Figure 7.1** depicts a healthy beehive that was suddenly encountered as the EP rounded a corner of the building. **Figure 7.2** shows a citrus canker decontamination station. In some instances, the EP may have to go through a decontamination zone after inspecting a property that is under quarantine. Some RECs such as an aboveground storage tank are readily observable and do not present any health or safety hazards to the EP inspecting the site. The complexity of the safety issues generally increases when the investigation of the RECs results in a health and safety hazard. The hazards also tend to increase with the level of the assessment. For example, in Phase II ESAs, the EP is actually trying to confirm contamination at the site and is trying to locate and sample the areas with highest concentrations.

In this chapter, we will discuss safety and health issues as they relate to Phase I/II ESAs. Our aim is to introduce issues for consideration that will be useful when planning and conducting any site investigation, regardless of type and duration. Basic training requirements under federal regulations, standard operating procedures, site safety plans, and initial site survey and reconnaissance will be discussed.

Potential exposure to toxic or chemically active substances during any type of environmental site investigation presents a special concern because they can be inhaled, absorbed through the skin, ingested, or directly introduced into our bodies through cuts or abrasions. Effects can vary depending upon the dose, contact time, type of substance, and lethality. As we already pointed out, site hazards are not limited to chemicals. Site hazards may include physical hazards (slip, trip, fall, electrical shock, falling objects, etc.), biological and radioactive material, heat stress, ergonomic factors, and many others.

There are two types of exposures that may result during on-site investigation. These are acute and chronic exposures. As a rule, acute exposures are of relatively short duration (minutes to 1–2 days), and the person is quite aware that an exposure has occurred because of the symptoms that appear. Chronic exposures are considered more "insidious;" that is, the exposure is continuous at low doses over a long period of time (months to years), and the person being exposed may not be aware that harm is being done until the symptoms appear. Because of the relatively short nature of Phase I and II ESAs, chronic exposures are not frequently encountered during these investigations. At any specific incident or site investigation, the hazardous properties of the materials may only represent potential risks whereas at other sites the risks are confirmed and significant. The health and safety of investigation personnel require that the risks be assessed and that appropriate measures be taken to minimize or eliminate the threat of exposure.

7.2 Training Requirements

Depending on the type and duration of the site investigation, many diverse types of health and safety training may be required under several different agencies. **Figure 7.3** lists some of the regulated training requirements that the EPs may have to undergo before performing certain portions of an ESA.

What is frequently lost in all of the regulatory mandated training is the simple fact that all mandates were designed to

FIG. 7.1   Beehive on the side of a building.

Source: Provided by Derrek Verlaan.

FIG. 7.2   Citrus canker decontamination zone. During the citrus canker eradication program, persons and equipment leaving areas with canker were required to go through decontamination.

Source: Photos provided by Derrek Verlaan.

FIG. 7.3 Training requirements [3,4].

Types of Potential Training Requirements

- *Contingency and Emergency Response Planning (OSHA, EPA and DOT)*
- *Confined Space Entry (29 CFR 1910.146)*
- *Hearing Protection (29 CFR 1910.95)*
- *Ergonomics (OSHA)*
- *Eye Protection (29 CFR 1910.132-133)*
- *Fire Prevention (29 CFR 1910.38)*
- *First Aid and Bloodborne Pathogens (29 CFR 1910.1030)*
- *Hand and Foot Protection (29 CFR 1910.132, 136)*
- *Hazard Communication (29 CFR 1910.1200)*
- *Lifting Techniques (OSHA)*
- *Lockout/Tagout (29 CFR 1910.147)*
- *Respiratory Protection 29 CFR 1910.134*
- *Tool Safety (CFR 1910.211-247)*
- *Environmental and Personnel Monitoring (EPA and OSHA)*
- *HAZMAT (Hazardous Materials) Shipping (49 CFR) DOT*
- *HAZWOPER (Hazardous Waste Operations and Emergency Response 29 CFR 1910.120)*
- *RCRA (Resource Conservation Recovery Act) EPA (40 CFR)*
- *Other EPA (Environmental Protection Agency) regulations (40 CFR)*
- *Other OSHA (Occupational Safety and Health Administration) regulations (29 CFR)*
- *Other DOT (Department of Transportation) regulations (49 CFR)*
- *State and local agency requirements*

protect personnel. Many times EPs are so busy preparing site safety plans, conducting training, and filling out required documentation to support their actions that they simply forget that the intent is to protect themselves and others from an injury on the job. Safety is first and foremost the EPs' personal responsibility. EPs need to take full responsibility for their actions while conducting any site investigation. To let their own protection depend entirely upon others and the mandated regulations is foolhardy at best. The EPs should also always remember to protect themselves and others first and only then consider protecting the equipment and the environment.

The two regulations most likely utilized by EPs during a Phase I/II investigation are the Occupational Safety and Health Administration (OSHA)'s HAZWOPER Standard and the training requirements under the U.S. Environmental Protection Agency (EPA)'s Resource Conservation and Recovery Act (RCRA). These regulations contain general requirements for safety and health programs, site characterization and analysis, site control, training, medical surveillance, engineering controls, safe work practices, personal protective equipment, exposure monitoring, informational programs, material handling, decontamination, emergency procedures, illumination, sanitation, site excavation, emergency response procedures, and sampling protocols.

As with any site investigation, preplanning is always the key. Training is a critical step in ensuring that all of the planning is correctly utilized. Trying to determine if a respirator should have been worn and, if so, which type after removing the bung from a drum for sampling is too late.

Once procedures have been put in place and training has been completed, following through with procedures may save the EP's life. An individual at a major chemical company was recently killed because he failed to follow procedures for which he had been trained many times. After being splashed with what he thought was a minor amount of a chemical, he decided to bypass the emergency shower (use of the emergency shower was required) and proceeded instead to a locker room shower, where he collapsed and died. These types of incidents reinforce the need for each individual to take responsibility for their own safety and to follow the procedures for which they have been trained.

7.3 Standard Operating Procedures

There are many procedures that may be required while performing the various tasks associated with a site investigation that could potentially involve hazardous substances. These procedures may be safety, technical, administrative, or management oriented. All of these procedures are intended to provide uniform instructions for accomplishing specific tasks.

If not appropriately trained, then EPs should seek out the help of competent professionals familiar with safety and health issues that will help them develop standard operating procedures (SOPs). **Figure 7.4** lists SOP principles that should be considered for all site investigations.

FIG. 7.4 Principles of SOPs.

Principles of Standard Operating Procedures

- *Prepared in advance*
- *Prepared using the best available information concerning the site and potential hazards*
- *Field-tested, periodically reviewed, and revised when necessary*
- *In writing and easy to use*
- *Procedures read and understood by all personnel before site entry with copies of SOPs available on site at all times*
- *Personnel trained in procedures*
- *Sufficiently generic to fit most site situations and adaptable to site-specific issues*

For any given site investigation, it is recommended that SOPs be adapted to protect on-site personnel against the hazards present at the specific site and for the types of activities to be performed (i.e., Phase II sampling).

7.4 Site Safety Plans

Site safety plans are developed to establish policies and procedures for protecting the health and safety of site assessment personnel during all on-site operations. The plan contains information about the known or suspected hazards, routine and special procedures, and other instructions for protecting the site assessment workers. It should also identify emergency procedures for unexpected site conditions.

Figure 7.5 outlines the content and issues that should be included and considered in a site safety plan. As with any plan, modifications should be made where appropriate to satisfy the needs of the investigation. Depending on the tasks to be performed and the potential hazards encountered, flexibility is the rule.

7.5 Initial Site Survey and Reconnaissance

This is perhaps the most challenging part of any site investigation. It is not uncommon for the EP to enter a site with little or no idea of what types of hazards may be present. Therefore, before entering any site for the first time, the EP should collect as much information as possible concerning the type or degree of potential hazards and the risks that may exist. Because most ESAs are conducted with strict time constraints, the EP may have limited time to independently research hazards. The users, owners, and especially the occupants may be the best sources of safety information associated with their operation. In many instances, the occupants will have their own safety and health requirements that the EPs will be required to follow when entering the facilities. The EPs should seek out this information and be cognizant of occupant-imposed requirements.

FIG. 7.5 Components of a site safety plan.

- *Introduction*
- *Key Personnel*
- *Task Safety and Health Risk Analysis*
- *Personnel Training Requirements*
- *Personnel Protective Equipment (if any is required)*
- *Medical Surveillance (if any is required)*
- *Emergency Response/Contingency Plan*
- *Confined Space Entry Requirements*

The review of the site occupants' safety information is useful to the EPs for two main reasons. First, the EPs gain an understanding of potential hazards and can plan to take appropriate steps to protect themselves during site inspection and any site activities. Second, the safety information enables the EPs to gain an understanding of the current use of the property with respect to hazardous chemicals and potential contaminants of the subject property. If the occupants have appropriate plans and documentation in place, then it demonstrates their awareness of environmental issues. If the occupants' documentation is lacking, then their awareness of environmental requirements may be inadequate and some of the operational/disposal activities may have been inappropriate. Lack of awareness of environmental issues increases the potential for contamination on the subject property and is a warning to the EP not to rely on the occupants' information for personal safety.

The safety issues can become complicated when the EPs are conducting site assessments on abandoned properties or buildings where the structures present physical hazards and lack of information about the former operations makes the chemical safety preplanning difficult (see **Fig. 7.6**). In these instances, research into the industrial uses of chemicals and associated toxicology can be extremely helpful.

With the advances in information technology, EPs have several options that can make the chemical research relatively quick. Many of these resources are free of charge. There are also other resources that can be purchased. For example, Patty's Industrial Hygiene and Toxicology [5] is a ten-volume professional reference work covering occupational health and safety standards. It is available in an online version that contains all of the text, references, charts, and graphics. It also includes the National Institute of Occupational Safety and Health *Manual of Analytical Methods*, which is a compendium of analytical methods for air and biological samples. The electronic version enables efficient and easy searches. EPs who do not have current safety and toxicology information in their office libraries should make appropriate efforts to establish research capability online, at public libraries, or with libraries on college and university campuses.

Types of hazards encountered during an environmental site inspection may include organic and inorganic vapors, gases, radiation, oxygen deficiency or oxygen-enriched atmospheres, combustible gases, and biological or physical hazards. Personal protective equipment (PPE) should be selected as appropriate using the four HAZWOPER categories (Level A—highest, Level B, Level C, and Level D—least) and should be based on hazard and risk of exposure.

If decontamination procedures are necessary, then it should be assumed (initially) that all personnel and equipment leaving the "exclusion zone" (area of potential contamination) are grossly contaminated. Initial decontamination plans are based on worst-case situations, or assume that no information is available about the incident. However, specific conditions at the site or the presence of other information from off-site should be evaluated for modification of the plan. Issues such as the type of containment, the amount of contamination, the PPE level, the equipment type required to

FIG. 7.6 Fire significantly damaged this building, making entrance unsafe.

Source: Provided by J. Kilgore.

complete the task (drill rig, hand auger, etc.), and other factors should be considered.

7.6 Subcontractor and Invitee Safety

One of the key considerations with respect to safety is the safety of all people that may be on the site during the assessment. This is particularly important in Phase II assessments, when the consultants actually set up and conduct operations on the site. The site can easily become a multiemployer site with site employees, consultants' employees, subcontractors' employees, potential invitees, and the public. If any of these persons are hurt on the site, then the consultant may be exposed to significant liability for the injuries. The complexity of these issues will vary from site to site and with the scope of the Phase II ESA. Many of these issues are typically addressed in contractual agreements.

7.7 Summary

In this chapter, safety and health issues associated with site investigations were discussed. The importance of appropriate training that stresses the protection of personnel and not just following the regulations was emphasized. The two most likely regulations (HAZWOPER and RCRA) to be of interest were indicated.

The types and importance of SOPs were also discussed. Items for inclusion in site safety plans were provided for reference. Also discussed was the difficulty with initial site entry due to unknown hazards and risks. The main focus of any health and safety plan is to protect human health. The consultant performing operations on a site is responsible for the safety of others.

References

[1] ASTM Standard **E1527**, Standard Practice for Environmental Site Assessments: Phase I Environmental Site Assessment Process, *Annual Book of ASTM Standards*, ASTM International, West Conshohocken, PA, 2013.

[2] ASTM Standard **E1903**, Standard Practice for Environmental Site Assessments: Phase II Environmental Site Assessment Process, *Annual Book of ASTM Standards*, ASTM International, West Conshohocken, PA, 2011.

[3] Occupational Safety and Health Guidance Manual for Hazardous Waste Site Activities, Occupational Safety and Health Administration, Washington, DC, 1995.

[4] Standard Operating Safety Guides, U.S. Environmental Protection Agency, Washington, DC, 1988.

[5] Clayton, G. D., Clayton, F. E., Cralley, L. J., Cralley, L. V., Harris, R. L., and Bus, J. S., *Patty's Industrial Hygiene and Toxicology*, 6th ed., online, John Wiley & Sons, New York, 2014.

Chapter 8 | Interviews

8.1 Interviews with Owners and Occupants

E1527 [1] requires the environmental professional (EP) to conduct interviews with owners and occupants. The objective of the interviews is to develop information about the subject property and its activities. To accomplish this goal, the EPs should seek out the person who has this information. The owners of the property are not necessarily the persons most knowledgeable about the uses of the property. Some are absentee owners and may be out of state with limited knowledge about the everyday business activities on the property. For this reason, E1527 requires the EPs to make a reasonable effort to locate and interview the "key site manager."

8.1.1 KEY SITE MANAGER

The key site manager is defined as the person having good knowledge of the uses and physical characteristics of the property. It can be the user, owner, an employee of the owner, or anyone involved in the everyday management of activities associated with the subject property. If the property is an industrial plant, then this person could be an environmental manager or health and safety officer of the facility. E1527 requires that the owner be asked to identify the key site manager (see Section 10.5.1 of E1527). It is good practice for the EPs to request that the user ask the owner (if different from the user) to identify the key site manager and provide a phone number for such a person. This should be done in writing at the outset of the process and included together with the request for helpful documents.

8.1.2 USER ISSUES

It can be challenging to identify the appropriate key site manager. If the lender is ordering the Phase I ESA for a refinance, the lender's site contact may be the CFO or an absentee owner. If a buyer is requesting a Phase I ESA, the buyer's contact may be the seller's broker or other agent of the seller. E1527 places the burden of identifying the key site manager solely on the property owner. Putting the request in writing at the outset of the Phase I ESA is helpful because it documents the EP's effort to locate the person most knowledgeable about the property to interview.

8.1.3 PAST OWNERS, OPERATORS, AND OCCUPANTS

Section 10.5.4 of the standard requires the EP to interview past owners, operators, and occupants of the property who are likely to have material information regarding the potential for contamination. These persons may be difficult to reach. **Figure 8.1** depicts an example of a past owner of a gasoline station that the EP will very likely be unable to interview. The standard includes additional language that recognizes these potential difficulties. It limits the obligation to interview to the extent that the persons have been identified and that the information likely to be obtained is not duplicative of information already obtained from other sources.

8.1.4 ABANDONED PROPERTIES

In the case in which the subject property is abandoned and there is evidence of potential unauthorized uses of the abandoned property or evidence of uncontrolled access to the property, it is necessary to conduct interviews with one or more owners or occupants of neighboring or nearby properties. Note that the requirement to interview neighboring property owners is based on a two-part test:

1. The subject property is abandoned. The standard provides further guidance for this determination by quoting directly from the U.S. Environmental Protection Agency (EPA)'s All Appropriate Inquiries Final Rule (AAI) regulation, which states that a "property that can be presumed to be deserted, or an intent to relinquish possession or control can be inferred from the general disrepair or lack of activity thereon such that a reasonable person could believe that there was an intent on the part of the current owner to surrender rights to the property."

2. There is evidence of potential unauthorized uses of the abandoned property or evidence of uncontrolled access to the property.

In the course of typical commercial real estate transactions, this might apply in the case of a potential foreclosure in which a property has been vacated, a property owner cannot be located, or the property has not been secured to prevent uncontrolled or unauthorized access. **Figure 8.2a** represents a situation that meets only part of this test. This property has been abandoned but has

FIG. 8.1 The EP will not likely be able to interview these past gas station owners. However, the EP will likely be able to draw conclusions regarding the potential for releases from this historical activity without the need for an interview.

Source: From the archives of the Draper Historical Society.

FIG. 8.2A This photograph depicts a property that has potentially been abandoned; however, doors and windows have been secured to prevent uncontrolled access.

Source: Provided by J. Kilgore.

been secured to prevent unauthorized access. **Figure 8.2b** depicts a condition that would meet both criteria, assuming the property owner has in fact abandoned the property.

8.1.5 RELATIONSHIP AND TIMING ISSUES

When conducting the interviews, it is imperative for the EPs to recognize the importance of the relationships among the user, the owner, and the key site manager. The relationship can influence the individuals' motivation to respond truthfully to questions during interviews. For example, in the case of an owner applying for refinancing, the owner has a vested interest (the loan) in minimizing the number of recognized environmental conditions (RECs) identified by the EP on the subject property. The owner may go to great lengths to clean up any evidence of contamination before the site visit and during the interview try to mislead the EP. A seller of a property, who is not the user, may be similarly motivated to ensure smooth transfer of the property to the new owner. There are not always sinister motives behind the lack of complete information. **Figure 8.3** depicts four apparent vent lines from underground storage tank (UST) systems, but the key site manager had previously stated there had been no USTs on the property. Upon further inquiry, the key site manager had forgotten that several USTs had previously been removed from the property in the early 1980s (well before current UST regulations).

Sometimes the sellers are the Phase I Environmental Site Assessment (ESA) users. For example, a prudent manufacturer may be selling an older property to relocate its manufacturing plant to a new facility. Because the user is concerned about contamination of the subject property by the new owners, they choose to perform Phase I ESA to document that the property was not contaminated at the time it was sold. In this case, the interviewee is motivated to answer truthfully and is willing to provide any documentation to support the answers given.

Although it is advantageous to have all available information before the time of the interview, time constraints placed on the EP by the user sometimes make it impossible to conduct the site visit concurrently with the interviews. In addition, EPs may not have the time to collect and review all of the historical records by the time they conduct the site visit. **E1527** Section 10.4 leaves it up to the EPs' discretion whether to conduct the owner and occupant interviews before, during, or after the site visit, with two exceptions. Before the site visit, the EP must request helpful documents and ask the user, owner, or key site manager whether they know of any proceedings involving the property (see Sections 10.8 and 10.9 of **E1527**). "Helpful Documents" outlined in Section 10.8.1 of the standard consist of the following:

- ESA reports
- Environmental compliance audit reports
- Environmental permits (e.g., solid waste disposal permits, hazardous waste disposal permits, wastewater permits, National Pollutant Discharge Elimination System permits, underground injection permits)
- Registrations for underground and aboveground storage tanks
- Registrations for underground injection systems
- Material safety data sheets
- Community right-to-know plans
- Safety plans; preparedness and prevention plans; spill prevention, countermeasure, and control plans; facility response plans, etc.
- Reports regarding hydrogeologic conditions on the property or surrounding area
- Notices or other correspondence from any government agency relating to past or current violations of environmental laws with respect to the property or relating to environmental liens encumbering the property
- Hazardous waste generator notices or reports
- Geotechnical studies

FIG. 8.2B Conversely, this photograph depicts a property that has been abandoned and where extensive vandalism has occurred. In this case, there would be a requirement under the standard to interview one or more neighboring property owners.

Source: Provided by J. Kilgore.

FIG. 8.3 Vent lines observed during site reconnaissance.

- Risk assessments
- Recorded Activity and Use Limitations (AULs)

Figure 8.4 lists the required elements of **E1527** showing a desirable sequence of events to maximize the efficiency of the Phase I ESA process. The section numbers refer to the respective sections of the **E1527** Standard where the requirements are contained.

E1527 also requires the EP to make at least one attempt to meet with the key site manager at the site. In this manner, the key site manager can be present during the walkover and answer any questions the EP may have at the time. From a practical standpoint, having the key site manager present at the time of the site visit has a tendency to make the assessment process more efficient. Questions that come up during observations can be answered immediately, and the key site manager can provide access to locked buildings and warn the EPs of potential physical hazards associated with the inspection.

8.1.6 OCCUPANTS

In commercial real estate transactions involving residential properties, **E1527** does not require the EPs to interview residential occupants. On occasion, a residential occupant may have specialized knowledge about current or past historical properties and can provide first-hand helpful information. If such a person is identified through the normal process of the Phase I ESA, then the EP may interview and use the information. The standard does not require the EP to engage in any extra effort to determine whether such persons exist.

Not all of the occupants at the property need to be interviewed. **Figure 8.5** outlines **E1527** requirements.

8.1.7 CONTENT OF THE INTERVIEWS

E1527 leaves it up to the EP to develop specific questions to ask. The point of asking the questions is to identify current and historical uses of the subject property with respect to RECs. The EP has to identify the owners and occupants interviewed and must report on the duration of their occupancy. The interviews can be conducted in person, by phone, or in writing.

Just as it was important to develop a standardized methodology for site reconnaissance, it is equally important to develop several questions to ask of the owners and occupants. **E1527** provides very little guidance on what questions the EP should ask. This is one reason why the EP generally develops a list of questions to cover during the interviews. As a minimum, EPs' questioning

FIG. 8.4 Chart of preferred sequence of events.

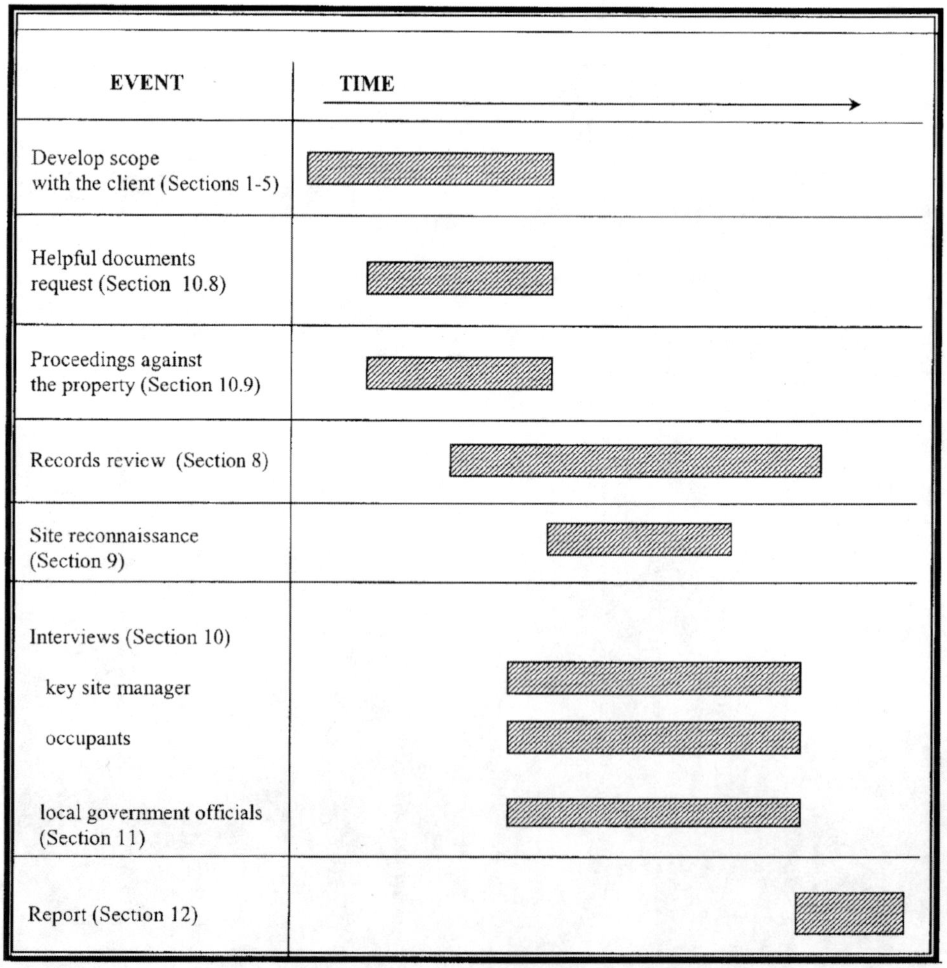

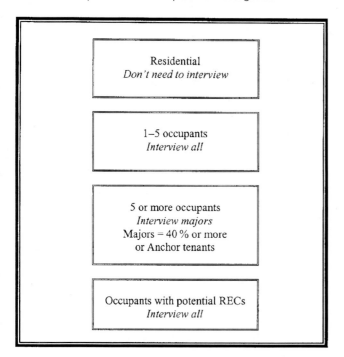

FIG. 8.5 Occupant interview requirement categories.

should cover the areas addressed in the site reconnaissance of the standard and any helpful documents that were obtained or became available during the interview.

The **E1528** [2] Transaction Screen questionnaire was primarily developed for the person with limited awareness of environmental issues associated with an ESA. It has recently been modified as a document for assessing business risk. Some EPs use the questionnaire as a guidance document for interviews. This may be most helpful to those initially learning to conduct interviews. The questions are very basic if used in their yes/no/unknown format. For the EPs to provide opinions, they generally seek more than just the yes/no answers. If using the Transaction Screen questionnaire, then the EPs should ask numerous follow-up questions to gain as much information on the particular subject as possible. The Transaction Screen questionnaire also serves as a documentation tool. Because it has a separate column for conditions observed at the site, it also enables direct comparison between interview findings and observed conditions at the site. Depending on the local requirements and needs of the EPs, Transaction Screen questionnaires are often customized by the EPs to include those issues.

To conform to **E1527**, the EPs are not required to conduct the interviews. However, if it is not the EP, then **E1527** requires in Section 7.5.1 that the persons conducting the interviews have sufficient training and experience and have the ability to identify issues relevant to RECs. In other words, they do not have to meet the AAI requirements for an EP, but they must have sufficient experience, and they must ask the questions and make reasonable attempts at providing a sufficient explanation of the meaning of the questions to the persons being interviewed. Persons being interviewed may have difficulty understanding the EP's jargon and may assume different meaning for the terms used. The guide sections of **E1528** provide some additional helpful information and further elaborate on the meaning of the questions. At a minimum, the EP conducting the Phase I ESA interviews should be familiar with the Transaction Screen guide. **E1527** assumes that the EP has sufficient training and experience to conduct the interview at a higher level of inquiry than presented in **E1528**.

With the exception of the user, the interviewee is not under any obligation to answer the EP's questions. The interviewee may also not know the answer to the questions. The EP should request that the persons being interviewed answer the questions in good faith and to the best of their knowledge. If the answer to a question is unknown or if the person refuses to answer the question, then it does not mean that the EPs failed to conform to the standard. The EPs need to provide adequate documentation of the questions asked and the responses received. If the questions were asked in writing, sent by mail, and no response was received, then the EPs have to attempt to follow-up at least once to conform to **E1527**.

It is often helpful to let the persons being interviewed know that the EP is seeking actual knowledge and information that can be independently verified. "Actual knowledge" is simply something that the individual knows. It does not include information that may be known by an entity such as their employer. Collecting second-hand information or hearsay is not the point of the interviewing process and is not actual knowledge. Letting the interviewee know at the beginning of the interview about these concepts helps to streamline the process. EPs should document all sources of the information provided.

Interviewing skills vary from one interviewer to the next. The skill is important for identifying RECs associated with the subject property. The interviewer must be a good listener to be able to follow up on key issues. For instance, if the occupant says that he has not seen any staining since the new floor was poured, then it should tell the interviewer that there was staining before the new floor was installed. The follow-up question would then address that issue.

Sometimes the EP is performing the Phase I ESA on a piece of property that previously had a Phase I ESA done. If the EP has a copy of the previous assessment including the information from the interviews of the occupants, then it is not necessary to ask the same occupant the same questions asked in the previous ESA. The EP can use the information, but they must inquire about any new information since the last interview.

8.2 Interviews with State and Local Government Officials

In addition to interviewing owners and occupants, the EPs must also interview one state or local agency official to comply with the minimum requirements of the standard (see Section 11 of **E1527**). **E1527** gives the EP the following choices:

- Local fire department serving the property
- Local or state health agency serving the property

- Local or state environmental or hazardous materials agency serving the area including the property
- Local agencies responsible for the issuance of building permits or groundwater use permits that document the presence of AULs, which may identify a REC in the area in which the property is located

Interviews with the local fire department can sometimes provide useful information with respect to the subject property. Because of the emergency management and response requirements, some fire departments collect a wealth of useful information about the businesses that they serve. They are frequently willing to share the information with the EP. Occasionally, the local fire department has no useful information or the officials are not interested in answering any questions.

Agency officials can be very helpful, but they are often too busy to spend time on interviews with the EPs. It is important to have specific questions in mind. The interview can be conducted as a follow-up on information obtained through records research. The EPs are often permitted to review the file and ask specific questions about information associated with the file.

The interviews can be done in person, by telephone, or in writing. The agency officials are under no obligation to answer the questions or even grant the interview. To comply with the standard, the EP must keep written records of attempted interviews along with records of the persons interviewed and their responses.

8.3 Summary

In this chapter, we learned about the requirements of **E1527** with respect to conducting interviews. We talked about the timing and coordination of the requirements in a manner that enables the EPs to maximize the efficiency of the process. We discussed the applicability of the **E1528** Transaction Screen process in assisting the EPs in developing a systematic approach toward the interviewing and documentation process. The need to conduct interviews with knowledgeable state or local government officials was also discussed.

References

[1] ASTM Standard **E1527**, Standard Practice for Environmental Site Assessments: Phase I Environmental Site Assessment Process, *Annual Book of ASTM Standards*, ASTM International, West Conshohocken, PA, 2013.

[2] ASTM Standard **E1528**, Standard Practice for Limited Environmental Due Diligence: Transaction Screen Process, *Annual Book of ASTM Standards*, ASTM International, West Conshohocken, PA, 2014.

Chapter 9 | Report Preparation

9.1 Evaluation of Data

We are close to finishing the Phase I Environmental Site Assessment (ESA) process. We completed the required elements of the scope of **E1527** [1], including the records research, inspecting the site, and conducting the interviews. We are now ready to evaluate the data and prepare the report for our client.

Section 12 of **E1527** is titled "Evaluation and Report Preparation." However, this section does not explain the evaluation process to the environmental professional (EP) as the title suggests. Instead, it primarily addresses the requirements with respect to the report preparation. Evaluation is an important element of the Phase I ESA and is usually the reason why the user hired the EP in the first place. Evaluation is the part of the Phase I ESA process in which the EPs interpret the data, applying their expertise to the facts gathered throughout the assessment process. It is often included as the opinion, analysis, or discussion section of the report.

Up to this point, the EP has been collecting information from several different sources. All of the collected information existed at the time the Phase I ESA was conducted and was "reasonably ascertainable," as defined by the standard. No new information was generated, no samples were taken, and the investigation into the recognized environmental conditions (RECs) at the site consisted of visual and physical observation of the environmental conditions associated with the subject property. Evaluation analyzes all of the assembled information together in preparation for rendering of the opinions with respect to RECs or the suspected RECs identified for the subject property.

9.2 Report Format

There is no required format for the report. Appendix X-4 of **E1527-13** provides a general description for all of the required elements of the standard and is a good place to start when preparing an outline for the specific components that will be included in the EP's report. Users may request that a specific format be followed. This is especially true of large banks and similar users who review a lot of reports and prefer a standard format.

9.2.1 DOCUMENTATION

This is an area in which many reports are deficient. The report should contain sufficient documentation either in the report itself or by reference to allow reconstruction of the work by an EP other than the EP who conducted it. Sources that revealed no findings also need to be included. Documentation includes scaled figures with north arrows and significant property features, photographic documentation if warranted for inclusion, and key historical research documents with the site location indicated.

9.2.2 CONTENTS OF THE REPORT

This section contains two important requirements. The standard is not specific as to where the information is included, but it does require that the report identify the EP in responsible charge of the assessment and the person who conducted the site reconnaissance and the interviews if they are different. It also requires the EP to state whether the user(s) provided the EP with information pursuant to the user responsibilities in Section 6. This is essentially the information contained in the optional "User Questionnaire" in **E1527-13** Appendix X-3 and includes any information about specialized knowledge. If the user fails to respond, then it will constitute a data gap. How the EP handles data gaps is discussed later in this section.

This active participation of the user is a significant change in the Phase I process. Historically, if asked, users have not provided any information or simply indicated that they have no information. Although it is true that the buyer has limited information about the property, the EPs need to document that they made a good faith effort to obtain and provide the information. As noted in Appendix X-3, "failure to provide this information could result in a determination that AAI is not complete." There is a difference between failure to respond and responding but not providing any useful information. There is no requirement to provide specific information. The requirement is to make a good faith effort to obtain it.

9.2.3 SCOPE OF SERVICES

The report should describe the scope of services. Some EPs do this by including a copy of the signed proposal (without cost information) in the Appendix. Others provide a brief summary or make reference to a specific proposal or authorization. It is very important to clearly identify the specific version of the ASTM Standard followed.

If any nonscope issues are addressed or if the user had requested recommendations, then the specific scope of work should be described.

For example, the client may want to assess the business risk associated with owning a building with asbestos. The presence of asbestos-containing building material being used for its intended purpose is not included in an ASTM E1527 scope. A statement regarding business risk of asbestos is not a specific scope of work. The scope should indicate whether it is a simple visual inspection for suspected friable materials or an actual survey where samples were to be collected for analysis. In the latter case, the scope would go on to describe the type of samples (friable or nonfriable or both), the location (inside or outside the structure or both), and the type of analysis to be run on the collected samples.

Some EPs include nonscope considerations as part of their Phase I ESAs whether requested by the client or not. This practice is strongly discouraged and can lead to liability exposure for the EP. There are any number of issues beyond the scope of E1527. Section 13 of E1527 identifies several possible "nonscope" issues, but this list is in no way exhaustive. Once the EP has "opened the door" to issues beyond the scope of the E1527, this provides a basis for the argument that because the EP included one issue (e.g., asbestos-containing material), then the EP should have informed the user of the potential for accumulation of radon in the lower level apartments. Providing additional services in conjunction with a Phase I ESA is fine. However, it is recommended that it is clearly identified in the report what scope of work was requested and what scope of work is beyond the E1527 and to clearly state that no other nonscope issues were included in the assessment beyond those specifically requested (or authorized) by the user.

9.2.4 FINDINGS

The findings section is a mandatory section. "Findings" are the facts uncovered throughout the environmental assessment process of E1527. Typically, findings are those items which are RECs, controlled recognized environmental conditions (CRECs), historical recognized environmental conditions (HRECs), or environmental concerns that may not rise to the level of the prior conditions. The property's current use as a distributing warehouse of roofing materials is a finding. Historical use of the property for agricultural purposes is also a finding. The findings should be documented in the report. It is possible that some of the findings may not provide any useful information with respect to the RECs associated with the property.

In the report, the EPs should include all of the findings that relate to the evaluation of determining whether conditions are ultimately determined to be RECs.

9.2.5 OPINIONS

Opinions on all findings must be provided. It is possible to handle them as groups. For example, there were two Comprehensive Environmental Response, Compensation, and Liability Act (CERCLA) sites and five leaking underground storage tank (LUST) sites identified in the government records within approximate minimum search distances, but they are all downgradient and would not be expected to affect the subject property. It is also common that multiple sources of information will identify the same item. For example, there is a record of a tank release on site, the owner indicated in the interview that there was a tank problem, the local health department representative stated that there was a reported release, and the EP observed the fill port and vent associated with the underground storage tank (UST). The sources of all of this information would be documented in each of the appropriate sections of the report and then can be summarized in the findings and opinions section by stating that there were multiple sources of information indicating there had been a release from a UST on the site. The report would go on to identify it as a REC because of the likely release to the environment.

Although there is a requirement for an opinion, there is no requirement for a separate opinion section in the report. Many EPs include the opinion with either the findings or the conclusions. For complicated sites with extensive findings, it may be easier to follow the discussion if there is an actual "opinion" section. For sites without complex issues or significant findings, the opinion discussion is easily combined with the findings or conclusions section.

Where the opinions appear in the report is less important than how the opinion is articulated. E1527 requires that the opinion discussion must include the logic and reasoning used by the EP in determining that a finding is or is not a REC. This does not have to be an extensive discussion, but it needs to be clear. In the previous examples, the CERCLA and LUST database listings were determined by the EP to not constitute RECs because of their downgradient locations. The EP determined that the on-site tank release was a REC because there were multiple indications of a past release.

There are some situations in which EPs may differ in their opinions regarding calling something a REC. It is especially important in those situations to be clear about the logic. An example would be a well-documented temporary storage of thousands of gallons of hazardous waste on a site with no indication of a past or present release. One EP may feel that the presence of that much hazardous material in an environment where hazardous material is being routinely handled, where the condition of the containers could not be verified, and where there is limited containment in the event of an accident constitutes a material threat of a future release and elects to call it a REC. Another EP might note the presence of a large volume of hazardous material but note that the operation has a good safety record and that all of the hazardous material is kept in sound containers and that there was no evidence of a past or present release. Therefore, they did not view the presence as a material threat of a future release, and it did not rise to the level of a REC.

The concept of an HREC was initially introduced in the 2000 version of E1527 and was revamped in the 2013 version. The HREC definition was developed to cover situations in which there has been an identified release that had been addressed. Up until the HREC was introduced, a remediated site would have met the definition of a REC because of the "past release." There is nothing in the REC definition that qualifies the current status of past contamination. The HREC provides a way of dealing with such situations. **Figure 9.1** presents the HREC as it is defined in **E1527-13**. Note that once an HREC has been identified, it is a finding and the EP must determine if it represents a REC currently. In a situation in which there had been a surface spill that had been excavated and removed with closure samples indicating no residual contamination, the conclusion could be that the HREC is not a REC.

REPORT PREPARATION

FIG. 9.1   Definition of HREC.

> *Historical recognized environmental condition:* A past release of any hazardous substances or petroleum products that has occurred in connection with the property and has been addressed to the satisfaction of the applicable regulatory authority or meeting unrestricted use criteria established by a regulatory authority, without subjecting the property to any required controls (for example, property use restrictions, activity and use limitations, institutional controls, or engineering controls). Before calling the past release a historical recognized environmental condition, the environmental professional must determine whether the past release is a recognized environmental condition at the time the Phase I Environmental Site Assessment is conducted (for example, if there has been a change in the regulatory criteria). If the EP considers the past release to be a recognized environmental condition at the time the Phase I ESA is conducted, then the condition shall be included in the conclusions section of the report as a recognized environmental condition.

Source: From ASTM **1527-13**.

FIG. 9.2   Definition of CREC.

> *Controlled recognized environmental condition:* A *recognized environmental condition* resulting from a past *release* of *hazardous substances* or *petroleum products* that has been addressed to the satisfaction of the applicable regulatory authority (for example, as evidenced by the issuance of a no further action letter or equivalent, or meeting risk-based criteria established by regulatory authority), with *hazardous substances* or *petroleum products* allowed to remain in place subject to the implementation of required controls (for example, *property* use restrictions, *activity and use limitations*, *institutional controls*, or *engineering controls*). A condition considered by the *environmental professional* to be a *controlled recognized environmental condition* shall be listed in the findings section of the *Phase I Environmental Site Assessment report* and as a *recognized environmental condition* in the conclusions section of the *Phase I Environmental Site Assessment report*.

Source: From ASTM **1527-13**.

Another example would be a tank closure with a no further action (NFA) letter from the regulatory agency. Additional facts are that it was a used oil release site closed in 1996, and although samples show low enough levels of petroleum hydrocarbons to issue an NFA letter for unrestricted use, no samples were collected for chlorinated solvents. The EP might conclude that this HREC is a REC because of the likely presence of chlorinated solvents from a past used oil release.

In the 2013 revisions to **E1527**, the CREC was a term developed to apply to conditions in which a past release has occurred and has been addressed using "risk-based" standards.

Figure 9.2 presents the CREC definition found in **E1527-13**. The CREC differs from the HREC in two very specific ways. First, the CREC applies to a past release that has been satisfactorily addressed, but some amount of residual contamination remains and there are certain controls that allow the contamination to remain. Second, the CREC meets the test of a REC due to the "presence or likely presence" of the contamination. Therefore, by definition, a CREC is a subset of a REC and must be included in the conclusions section of the report.

An example of a common CREC that has been incorporated in the **E1527** definition is that of a LUST site that has been cleaned up to a commercial use standard and does not meet unrestricted residential cleanup criteria. This condition would be considered a CREC because the "control" is represented by the restriction that the property use remains commercial.

It is important to note that a condition identified by the EP as a CREC does not in any way imply that the EP has evaluated or confirmed the adequacy, implementation, or continued effectiveness of the "control," but rather it is an acknowledgment that there has been a release, the release has been addressed, but it is okay for the residual contamination to remain so long as certain conditions are met.

CRECs are very common, and under the 2013 version of **E1527** there will be far more conditions identified in the Conclusions section of the report than have previously been listed. In the past, EPs may have considered a risk-based closure to be an HREC, de minimis, or a business environmental risk. The 2013 revision has made it very clear that any site that has been addressed to anything less than an unrestricted residential standard must be identified as a CREC and must be included in the Conclusions section of the report as a REC. This will be a shift for users and EPs who have been conditioned to avoid having anything identified in the Conclusions section of the report. However, it is the unsuspecting day care in a former thermostat factory, the garden-level apartments in the old paint manufacturing building, or the excavation contractor unknowingly giving away contaminated soil as "free fill dirt" that have found themselves in trouble because they had no idea there were continuing obligations tied to that NFA letter.

As can be seen from the above discussion, it is in the opinion discussion that the reader can obtain an understanding of how the EP evaluated the information and arrived at his or her conclusion that a condition is or is not a REC. In many cases, the RECs or lack of RECs are obvious. "Obvious" in this case means that virtually all EPs would come to the same conclusion. There is an intermediate area where EPs may differ in their opinions. Both can be in full compliance with U.S. Environmental Protection Agency (EPA) All Appropriate Inquiries Final Rule (AAI) [2] and ASTM. There is no bright line in the standard. If the opinion is later challenged, then the discussion will focus on the logic and reasoning that the EP used to develop their conclusion.

Opinions should never be guarantees or recommendations. A statement that a particular finding did not affect the subject property can be interpreted as a guarantee. Unless such a statement can be substantiated with sufficient findings, it should not be made.

In the mid-1980s, ASTM published several standards aimed at product liability matters. Although product liability is not necessarily directly related to the environmental assessments, these standards outline appropriate scientific principles that should be part of any scientifically based opinion. **E678** [3] is a *Standard Practice for Evaluation of Technical Data*, and it addresses the issues that constitute a proper basis for formation of technical opinions. It deals with hypotheses and opinions based on consideration and analysis of technical data. **E620** [4] is a *Standard Practice for Reporting Opinions of Technical Experts*. The standard covers the scope of information to be contained in written technical reports that expresses an opinion about the adequacy or inadequacy of products that are or may be reasonably expected to be the subject of litigation. Many of the points required by the standard under the headings of "Descriptive Information" and "Pertinent Facts" are relevant to the work of the EP.

These ASTM Standards provide a scientific framework that should be a part of any expert's opinion, whether it is in the product litigation field or in the environmental field. After all, the EPs are experts with respect to the environment and are bound by the same scientific principles.

9.2.6 ADDITIONAL INVESTIGATIONS

This language in the standard is derived from language contained in the EPA's AAI regulation. The AAI language simply states, "The inquiry of the environmental professional should include an opinion regarding additional appropriate investigation, if any."

To fully understand this language, we need to look at the source and history of this language. It is tied to the AAI discussion about the obviousness of the contamination. This language is not new and has always been part of CERCLA. The preamble to the AAI makes several important points.

One is that EPA did not intend to change the way in which the Phase I studies are conducted. That is to say that EPA is not imposing new requirements or altering the practice. Both AAI and the ASTM Standard in Section 12.6.1 indicate that this opinion is not intended to constitute a requirement for Phase II recommendations or other assessment activities. It also indicates that the final rule does not require sampling and analysis as part of AAI.

The ASTM language provides more qualifying language. Looking at the ASTM language (**Fig. 9.3**), it would suggest that these are unusual circumstances when three conditions apply:

1. There is a REC.
2. More certainty is required.
3. The EP has an opinion regarding appropriate additional investigation.

This is subject to interpretation. EPA uses two examples in the preamble to the rule. One of those is a circumstance when it might be appropriate to conduct additional investigation either pre- or postacquisition to fully understand the conditions at the property to comply with the statutory requirements for CERCLA liability protections associated with postacquisition continuing obligations. The second example was when additional investigation would be required to address existing significant data gaps.

9.2.7 DATA GAPS

The standard requires the EP to comment on significant data gaps that affect the EP's ability to identify RECs. The standard also

FIG. 9.3 ASTM description of additional investigation.

> *Additional Investigation*—The *environmental professional* should provide an opinion regarding additional appropriate investigation, if any, to detect the presence of *hazardous substances* or *petroleum products*. This opinion should be provided in the unusual circumstance when greater certainty is required regarding the identified recognized environmental conditions. *A Phase I Environmental Site Assessment* which includes such an opinion by the *environmental professional* does not render the assessment incomplete. This opinion is not intended to constitute a requirement that the *environmental professional* include any recommendations for Phase II or other assessment activities.

Source: From ASTM **1527-13**.

FIG. 9.4 Definition of data gap.

> *Data gap:* A lack of or inability to obtain information required by this practice despite good faith efforts by the environmental professional to gather such information. Data gaps may result from incompleteness in any of the activities required by this practice, including, but not limited to, site reconnaissance (for example, an inability to conduct the site visit) and interviews (for example, an inability to interview the key site manager, regulatory officials, etc.).

Source: From ASTM **1527-13**.

requires the EP to identify other sources of information consulted to resolve data gaps.

Data gaps are defined in **Fig. 9.4** and will typically arise from various limitations in the Phase I process such as site visit limitations (discussed in **Chapter 6**), data failure in historical research (discussed in **Chapter 5**), and client-imposed constraints (interview restrictions).

The one important clarification of the AAI concept in the ASTM Standard is that the data gap must be significant. Simply having a data gap does not require comment by the EP. The standard goes on to give some examples of "significant." It further qualifies "significant" by indicating that a data gap would only be significant if the EP's professional experience or other information raises reasonable concerns.

This qualifier helps clean up an ambiguity in AAI. How does one comment on the importance of something you do not know? You may have a data gap because of an inability to find information (history, records, interviews) about the property use before 1920. If there is no other indication that the lack of information may be important, then it is not a significant data gap. You do not need to engage in speculation on what might or might not have happened on that property before 1920.

However, if the same area of the site had a confirmed history of industrial use and the site was identified as a foundry in 1880, the lack of information on the uses between 1880 and 1920 would be a significant data gap.

Or perhaps the subject site was identified on the regulatory database as a LUST site and the agency files could not be obtained within the "reasonable time" parameters of **E1527**. The EP may have requested documentation from the fire department, health department, and the property owner but still was unable to obtain documentation regarding the release. This would be another example of a significant data gap.

9.2.8 CONCLUSIONS

EPs evaluate findings and determine whether any constitute a REC. **E1527** requires that the EP identify, in the Conclusion section of the report, any findings the EP decides are a REC. For example, if the EP concludes that a 30-year-old hydraulic lift with an underground tank may have had a release to the environment, this constitutes a REC and must be stated in the conclusions.

The decision to identify certain findings as RECs will depend to a large extent on the EP's judgment, the EP's risk tolerance, and the EP's interpretation of the **E1527** definition of a REC. At a minimum, the EPs should consider how defendable their particular interpretation of the standard's language is. How would the EP defend the conclusion if they were questioned by other EPs, users, owners, or their attorneys?

The Conclusions section is one of the few places in the standard where specific language is required. If no RECs were identified, then a statement substantially similar to the language in **Fig. 9.5** is used. If RECs are identified, then the language in **Fig. 9.6** is used.

The opinions include the EP's opinion on the effect of various identified conditions or findings. There is no specific requirement about where that discussion should take place. This discussion could be provided in the Conclusion section. Note that there is a distinction between discussion of the logic that led to the opinion that something is a REC and the opinion of the EP on the effect of the REC on the property. An example of the latter might be the conclusion that contaminated groundwater from historical industrial activity is suspected or known to be present beneath the target property; however, groundwater is more than 200 ft deep and culinary water is provided by other sources. Therefore, the presence of the contamination would constitute a REC, but the effect of the REC on the property is minimal given the EP's understanding of current conditions and uses.

There is a tendency on the part of users and other reviewers of Phase I reports to flip to the back of the report and read the conclusions to see if any RECs were identified. For reasons discussed above, it is equally important for users or reviewers to read and understand the findings and opinions discussion in the report.

9.2.9 ADDITIONAL SERVICES

The standard simply acknowledges that there may be additional services associated with the Phase I. These services generally fall into two categories. One is expanded information about Phase I related issues. This could include risk assessments, more detailed conclusions, or recommendations. The second type is discussed in nonscope considerations. These would be things such as radon, asbestos-containing building material, and lead-based paint assessments.

E1527 does not require the EP to provide the user with recommendations. We have already mentioned that the opinion about additional services should not be construed as a requirement for the EP to provide recommendations. In Section 12.15, recommendations are specifically listed as beyond the scope (see **Fig. 9.7**).

If the EPs provide recommendations without users' requests, then they are opening themselves up to potential legal action because they are going beyond the scope of the standard. There is a logical reason behind the exclusion of recommendations from the scope. Presumably, the EP is an expert and, from an environmental standpoint, is a lot more qualified to make recommendations than the user. If the user does not follow the EP's recommendations, then they have practically no defense in court. The standard specifies that the user must decide what to do about the REC. By making the recommendation, the EP has stripped the user of his ability to apply business judgment to the issues of the transaction. When performing Phase I ESAs in accordance with **E1527**, the EP should not make recommendations unless specifically contracted to do so by the user. If requested, recommendations should be based on the user's business objectives and risk tolerance, which need to be discussed with and fully understood by the EP.

The EPs should also take care not to inadvertently provide recommendations as part of their opinions. A statement that a REC appears to have affected the property and should be

FIG. 9.5   ASTM required language—no RECs.

"We have performed a Phase I Environmental Site Assessment in conformance with the scope and limitations of ASTM Practice **E1527** of [insert address or legal description], the property. Any exceptions to, or deletions from, this practice are described in Section [ ] of this report. This assessment has revealed no evidence of recognized environmental conditions in connection with the property,"

FIG. 9.6   ASTM required language if RECs.

"We have performed a Phase I Environmental Site Assessment in conformance with the scope and limitations of ASTM Practice **E1527** of [insert address or legal description], the property. Any exceptions to, or deletions from, this practice are described in Section [ ] of this report. This assessment has revealed no evidence of recognized environmental conditions in connection with the property except for the following: (list)."

FIG. 9.7   ASTM required language regarding recommendations.

Recommendations: Recommendations are not required by this standard. A *user* should consider whether recommendations for additional inquiries or other services are desired. Recommendations are an additional service that may be useful in the *user*'s analysis of LLPs or *business environmental risk*.

Source: ASTM **E1527-13**.

investigated further is an opinion that includes a recommendation. A statement that "to determine if there has been an impact, additional investigation would be needed" is not a recommendation.

One final aspect of Phase I recommendations is that they increase the liability potential of the EP. To make sound recommendations, the EP must have sufficient data to support them. Phase I ESAs seldom provide sufficient data to form a solid scientific basis for specific recommendations. The fewer data the EP has, the higher the risk of making a wrong recommendation.

9.2.10 DEVIATIONS

The standard requires a listing of all deviations, including those imposed by a third party. Note that deviations made by the EP are conscious decisions not to follow the standard. This should be done only in those circumstances in which there is a good reason. An example would be that none of the government records on the other side of the river were reviewed because they cannot directly affect the subject site.

This section should also contain a description of any user-imposed constraints. These might be things like a very short delivery date for the report that affected availability of records, limited or restricted access to the property, or directions not to interview certain parties who would typically be interviewed.

Lastly, this section should also include discussion of physical limitations. This would include issues such as snow covering the ground, steep ravines, or impenetrable vegetation on the site.

9.2.11 REFERENCES

The report should contain references to all sources used in preparation of the report. They should be properly annotated so that they could be retrieved by a third party. These references are frequently included in an appendix.

9.2.12 SIGNATURES

The EP responsible for the report must sign the report. It is not uncommon to include the signatures of others who worked on or reviewed the report. There is no specific requirement as to where the signatures appear in the report. The most common practice is to include a signature page after the cover page or sign at the end of the report.

9.2.13 EP STATEMENTS

There are two statements that need to be included in any report. **Figure 9.8** is a direct citation taken from AAI requirements. One is a statement regarding the qualifications of the EP. It is simply a statement that the EP meets the qualifications of **E1527** Appendix X-2. The second one is a general statement of the qualifications in regard to the type of property being assessed in the Phase I.

This language has liability implications. EPs need to consider that they will have to make these representations. This should be well thought out before accepting a client with a site that includes an unusual history of use or for industrial sites involving exotic materials when the EP has had limited prior experience with sites of similar nature, history, and setting.

FIG. 9.8 Required professional statements.

> 1. "[I, We] declare that, to the best of [my, our] professional knowledge and belief, [I, we] meet the definition of *Environmental professional* as defined in §312.10 of 40 CFR 312" and
> 2. "[I, We] have the specific qualifications based on education, training, and experience to assess a *property* of the nature, history, and setting of the subject *property*. [I, We] have developed and performed all the appropriate inquiries in conformance with the standards and practices set forth in 40 CFR Part 312."

Source: ASTM **E1527-13**.

9.2.14 APPENDICES

The Phase I ESA report seldom includes all of the EP's file materials. Some of the handwritten notes or recorded interviews may have been transcribed, and documentation that did not reveal any significant findings may not end up being included in the report. For example, only aerial photographs showing significant changes in the uses of the subject property may end up being included in the report, with the rest being maintained in general aerial photography files at the EP's office. The standard requires that the reports contain sufficient information to support the conclusions reached. The EPs need to decide what information is sufficient to meet that requirement. It is permissible to include the specific source information by reference.

The report must list the names and qualifications of the EPs involved in the Phase I ESA (typically in an appendix). The qualifications of the EPs responsible for the report, as well as the persons involved in the site reconnaissance and interviews (if conducted by someone other than the EP), must be included.

EPs also need to develop internal document retention policies. How long are the files maintained? Is there a point in time when only the essential documentation is retained and the rest is discarded? Who owns the data files or any photos, and how long should they be retained? Once these policies are established, it may be helpful to advise or disclose these policies to the user. User-specific business needs may impose special requirements, and they may assume that the file information will be accessible forever. **E1527** does not provide any guidance with respect to document retention policies.

Under certain conditions, the EP may not be able to finish the Phase I. For example, the EP is instructed that, if while performing the Phase I the EP encounters indications of a REC that may be serious, the EP is to promptly notify the user. The user terminates the Phase I ESA, does not want to pay for any further effort to complete the project, and does not want or need a report. The EP needs to have documentation and a retention policy for such situations.

Sometimes when the Phase I identifies a REC without sufficient evidence to quantify the impact, the user may want to go directly into a Phase II ESA and combine the two investigations in one report. Most Phase I ESAs are fixed cost projects, and it may be helpful to have an agreement with the user on partial compensation if some issue causes a termination of the Phase I ESA or if the scope has to be expanded before the completion of the report.

9.3 Summary

In this chapter, we discussed issues associated with preparation of Phase I ESA reports. We stressed the importance of a scientific approach to evaluation of data and reviewed the concepts of findings, conclusions, opinions, and recommendations. We touched on some of the required elements of **E1527**, including additional investigation, data gaps, deviations, and professional statements, and we discussed some of the documentation issues facing the EPs.

References

[1] ASTM Standard **E1527**, Standard Practice for Environmental Site Assessments: Phase I Environmental Site Assessment Process, *Annual Book of ASTM Standards*, ASTM International, West Conshohocken, PA, 2013.

[2] U.S. Environmental Protection Agency, "All Appropriate Inquiries," Final Rule 40 CFR Part 312—Innocent Landowners, Standards for Conducting All Appropriate Inquiries.

[3] ASTM Standard **E678**, Standard Practice for Evaluation of Technical Data, *Annual Book of ASTM Standards*, ASTM International, West Conshohocken, PA, 1984.

[4] ASTM Standard **E620**, Standard Practice for Reporting Opinions of Technical Experts, *Annual Book of ASTM Standards*, ASTM International, West Conshohocken, PA, 1985.

Chapter 10 | Dealing with the Nonscope Issues

10.1 Nonscope Considerations Introduction

In many instances, the user may need to conduct an investigation into certain environmental issues that are beyond the scope of **E1527** [1]. These may include radon, lead-based paint, lead in drinking water, asbestos-containing building materials, endangered species, wetlands, mold, regulatory compliance, or air quality issues. Users may realize some cost benefits if the environmental professional (EP) can combine these additional investigations with the Phase I Environmental Site Assessment (ESA). One site visit may be sufficient to cover requirements of the related investigations, and some of the other tasks may also be combined to increase the efficiency of the process.

Not all "nonscope" considerations are as clear cut as they might initially appear. Recall that a recognized environmental condition (REC) is grounded in CERCLA (Comprehensive Environmental Response, Compensation, and Liability Act) liability and must be related to "releases" to the "environment." Consider the following examples:

- Asbestos is not always a nonscope consideration. There are numerous CERCLA sites related to releases of asbestos to the environment. However, an evaluation of asbestos-containing building material or lead-based paint that may be present in a building that is being used for its intended purpose would be beyond the scope of a Phase I ESA.
- An evaluation of indoor air quality unrelated to a release of hazardous substances or petroleum products into the environment is not included within the scope of an **E1527** Phase I ESA. Examples would include an evaluation for mold or worker exposures from on-site operations. These issues may be material to a client, but they are not related to releases of hazardous substances or petroleum products to the environment; therefore, they would not fall within the scope of an **E1527** Phase I ESA.
- Releases from a substance that is in an unaltered form or is altered by natural processes and located where the substance is naturally found are also beyond the scope of an **E1527** Phase I ESA. The primary example of this nonscope item would be the potential for the presence of naturally occurring radon gas.

However, when the source material for the radon gas is radioactive waste material that has been disposed or spilled, the likely presence of the radon gas would be a "release" of a "hazardous substance" that would fall within the scope of an **E1527** Phase I ESA.

- There is a distinction between the issue of vapor "migration" and the issue of vapor "intrusion." Recall from **Chapter 5** that the EP must evaluate the potential for a release to migrate onto the subject property, and one of those mechanisms for migration is by way of vapor. The EP must consider vapor migration in much the same way as groundwater migration. Once the EP has concluded that the release of the hazardous substance or petroleum product is likely to be present at the property, the Phase I ESA requirement to identify the REC has been satisfied. An evaluation as to whether vapor has migrated into a building, or groundwater contamination has affected a drinking water well, or soil impacts extend to a utility corridor, etc., are not within the scope of a Phase I ESA. This is not because these issues are unrelated to a CERCLA release but because these evaluations are examples of additional investigation regarding the nature and extent of the release and not within the scope of a Phase I ESA.

The **E1527-13** legal appendix included as Appendix XI in the standard provides extensive background information regarding congressional background, EPA interpretations, and case law regarding issues that are within and beyond the scope of an **E1527** Phase I ESA. Because regulations change, statutes are amended, and case law varies widely, this legal appendix is only intended for informational purposes and is not intended to be nor interpreted as legal advice.

10.2 Other Standards and Regulations

Many of the nonscope issues may be covered by other standards or regulated by other federal, state, or local requirements. For example, the user may want to conduct an asbestos or mold assessment in conjunction with the Phase I ESA. The state may require that these assessments be performed in accordance with state regulation and a specific standard. The EP may also have to meet specialized qualifications or licensing requirements to perform these assessments. The EP should only perform assessments within the EP's qualifications and must be aware of applicable standards and regulations.

Because there are many other federal, state, and local environmental laws and regulations that fall outside of the scope of an **E1527** that may be of interest to a property transaction, the ASTM task group developed a new informational appendix, Business Environmental Risk (BER), for the **E1527-13** Standard. Appendices included in any ASTM Standard are nonbinding and are intended to provide additional information (such as the **E1527-13** Legal Appendix) or optional tools (such as the **E1527-13** Use Questionnaire). The BER appendix expands on Section 13 of the **E1527** Standard, illustrating that the **E1527** practice does not address all possible environmental liabilities that a user may need to consider in the context of a real estate transaction, and it provides information regarding additional resources that may be utilized to address some of the more common issues that are beyond the **E1527**.

If any of these other issues are going to be included with the **E1527** assessment, then the EPs should take care to clearly separate the nonscope investigations in the report. This is often accomplished by dedicating a specific section within the body of the report to these issues or by using an independent appendix section to address the nonscope portion of the work. The independent appendix or section then contains separate scope statements, findings, and conclusion sections. The EPs should not attempt to incorporate or combine multiple out-of-scope investigations in the main body of the **E1527** report even if they were conducted at the same time. The investigations may require elements common to both but different in purpose, and combining the issues in the same section of the report will create confusion.

Another way to handle the out-of-scope investigations is to write separate reports, each within the scope of the applicable standard. This is, by far, the simplest method to achieve clear separation between the tasks, but it may result in some administrative duplication of report sections, such as property description, historical research, or site photographs.

10.3 Special User Needs

In addition to nonscope issues governed by other standards and regulations discussed above, the user may have special needs that could be resolved during the Phase I ESA. For example, the user may need to find out what environmental issues may arise if after purchase he decides to demolish and remove certain structures. Depending on the EP's expertise, these needs may result in additional business opportunities to expand the scope of work during the Phase I ESA. The Phase I ESA investigation can include special components to collect information needed to meet user needs. When expanding the scope of the Phase I ESA, the EP and the user should take care to clearly identify details of additional services and resolve any contractual issues. The report should sufficiently describe any additional services in a separate section.

10.4 Recommendations

In **Chapter 9**, we noted that recommendations are not required by the standard and that the EP should never include them in the Phase I ESA report unless the contractual agreement specifically requested the EP to do so. Sometimes the users do not realize the implications associated with requesting recommendations in the Phase I ESA report. When the user wants recommendations, the EP should discuss this nonscope issue with the user and explain the potential implications of making recommendations on the outcome of the transaction, including the user's ability to make decisions. The user and the EP should agree on the manner and the medium in which the recommendations will be delivered. The following are some of the common methods used in the industry to deliver recommendations.

10.4.1 VERBAL RECOMMENDATIONS

Some users prefer verbal recommendations only. Under some conditions, they direct the EP to deliver the recommendations verbally to their attorney, and they want to be completely left out of the process. The reason behind involving the attorney is to minimize the amount of discoverable information. If the user is sued, then the attorney will claim the "attorney client privilege" with respect to what he and the user discussed. However, the discussion between the attorney and the EP is discoverable. For their own protection, the EPs must decide on the method of documentation of verbal communications between the EP and the client. In verbal recommendations, documentation can potentially be a very sensitive issue. The very reason why the clients want verbal recommendation is that it does not leave a paper trail. In litigation, if the EP made a note in his file that recommendations were given, then the entire EP file, including notes, is discoverable information and could be subpoenaed. If the EPs make recommendations and do not make a record of what they were, then they will not have any tangible evidence to prove that recommendations were made and what they were.

10.4.2 WRITTEN RECOMMENDATIONS

Written recommendations solve the documentation problems but they create a paper trail, and even if they are submitted as a separate document, they are readily discoverable if litigation arises. Written recommendations are either included in the main body of the report, included as an attachment, or they are submitted separately in the form of a letter. If the EP decides to include recommendations in the main body of the report or in an attachment to the report, then the scope section of the report should list this as a nonscope issue and describe the contractual items agreed upon with the client. If the written recommendations are submitted in a separate letter form, then the scope section of the Phase I ESA of the report does not have to discuss recommendations.

10.5 Summary

In this chapter, we talked about some of the nonscope issues associated with Phase I ESAs. We pointed out that many issues may be regulated by other standards or agencies or both. We looked at how tasks that may be specific to the user or to the transaction can be combined with but addressed separately in the report to minimize confusion. We also discussed choices of how recommendations can be provided to the users.

Reference

[1] ASTM Standard **E1527**, Standard Practice for Environmental Site Assessments: Phase I Environmental Site Assessment Process, *Annual Book of ASTM Standards*, ASTM International, West Conshohocken, PA, 2013.

Part 2

Phase II Environmental Site Assessment Process

Chapter 11 | Scope of Phase II Environmental Site Assessment

11.1   Introduction To Phase II ESA

In Part 1 of this manual, we concentrated on performing Phase I Environmental Site Assessments (ESAs) in accordance with ASTM **E1527** [1]. Part 2 of this manual focuses on some of the key aspects associated with performing the Phase II ESAs in accordance with ASTM **E1903** [2].

In 2011, the ASTM task group made significant revisions to the Phase II standard. The 1997 version of the Phase II standard [3] allowed great flexibility, enabling the user and the consultant to design the process to meet the needs of the user. The **E1903** used to be a standard guide. It was a compendium of information that identified many options and allowed the parties considerable leeway in the scope issues. ASTM standard guides do not require a specific course of action, and the consultant had to determine the appropriate course of action that met the goals established with the client. Many ASTM task group members believed this flexibility to be burdensome and that it actually diminished the usefulness of the standard. Because the standard was also being used in the regulatory environment, particularly in Brownfields investigations, several regulators and consultants also felt that a more structured approach was needed and suggested changes to make the Phase II ESA standard more prescriptive.

Through the consensus process, this new approach was adopted and in 2011 the **E1903** changes from standard guide to standard practice came into effect. **E1903** is now just like **E1527**, a standard practice. As such, it provides the parties involved with a definitive set of instructions for performing a specific set of tasks. The Phase II assessor, a newly defined term by the new standard practice, strives to complete the required tasks and in this manner fulfills the requirements of the standard practice. In this manual, we will use the term "consultant" and "Phase II assessor" interchangeably. Both the consultant and the user now have a defined set of choices that dictate a defined process and responsibilities. The manner in how the process met the standard must be disclosed in the Phase II ESA report.

The Phase II standard revisions include a description of the scientific method that is appropriate for the Phase II investigations. The Phase II assessment uses a scientific method to characterize the property conditions in an objective, representative, reproducible, and defensible manner. This was considered a significant improvement because many Phase II environmental investigations before the 2011 revision did not follow or ignored the basic principles of scientific investigations. Consequently, this led to the generation of inappropriate data and erroneous conclusions being formed about conditions of the property.

It is noteworthy that many Phase II ESAs are done without the involvement of a regulatory agency. It is typically the advent of contamination discovery during the Phase II assessment that triggers the regulatory involvement. The revised standard now forces the Phase II process to follow the minimum principles of scientific investigation. This enables the data to be more useful to the regulators if contamination is discovered.

Some of advantages of making the Phase II into a standard practice came at a price. In fact, as we will learn further in this manual, the **E1903** Standard Practice now requires submitting a written report. When **E1903** was a guide, a report was only prepared if the user requested it.

11.2   Scope of **E1903-11**

With the new standard practice, we will learn that user needs play an important role in Phase II ESAs and should drive the process.

The advent of the Brownfields Amendments, the subsequent EPA rule, and revisions of the Phase I standard provided several new applications for the Phase II process. Brownfields are about redevelopment and reuse of previously developed and potentially contaminated properties. The Phase II investigations can be and often are designed to assist users with redevelopment decisions. The Phase II assessment may also be a useful tool in the case of Bona Fide Prospective Purchaser (BFPP). In this situation, the Phase II assessment may not be limited to identifying contamination but used for an extensive investigation into the actions necessary to meet the user obligations under the U.S. Environmental Protection Agency (EPA) All Appropriate Inquiries Final Rule (AAI). The investigations typically associated with BFPP tend to be designed to identify as much historical contamination as feasible so that the new owner can minimize the potential of being responsible for contamination that they did not cause.

Comparing the Scope sections of the two standards, we should note that the Phase I ESA was focused on identifying recognized environmental conditions (RECs). The Phase II ESA is designed to evaluate the presence of substances including but not limited to those within the scope of the Comprehensive Environmental Response, Compensation, and Liability Act (CERCLA) pollutants, contaminants, petroleum and petroleum products, controlled substances, data gaps, and even Phase I nonscope issues such as asbestos or lead-based paint.

The interest in any of these issues may arise from a wide variety of legal, regulatory, and commercial concerns and may involve diverse objectives that can be beyond any of the conditions identified in the Phase I ESA.

A Phase II ESA may arise as a result of site assessment activities followed in the standards, such as

- Practice **E1527** for *Phase I Environmental Site Assessments: Phase I Environmental Site Assessment Process*
- Practice **E2247** for *Environmental Site Assessments: Phase I Environmental Site Assessment for Forestland or Rural Property*
- EPA's AAI, 40 C.F.R. Part 312
- Practice **E1528** for *Limited Environmental Due Diligence: Transaction Screen Process*

The Phase II ESA may also be conducted to answer questions that may not have been addressed by any of the previous assessments. It should be also noted that whether consultants have classified certain property conditions or areas as RECs or not RECs may be entirely irrelevant to the current Phase II ESA.

The consultant must work with the client to define the scope and objectives of the Phase II investigation in light of all of the relevant factors and select an assessment approach needed to achieve sufficient confidence to answer the questions posed in the investigation. The investigation may not be accomplished in a single step and may be designed to work through several iterations. This allows the user to terminate or continue the process based on the information that is uncovered during these iterations. **Figure 11.1** demonstrates the concept.

In **Fig. 11.1**, the shaded area on the left represents several of the potential issues that the Phase II ESA is attempting to resolve. These issues are used to define the scope of the Phase II ESA. In the right-hand side example in **Fig. 11.1**, the user and the consultant agreed on three iterations of the Phase II ESA investigation. At each level, the consultant and the user agreed to review the results, and the user would then decide whether to proceed to the next level. To be able to assist the user in the development of the Phase II

FIG. 11.2   Conceptual representation of the relationship of the standards, the tasks, and the conceptual site model (CSM) thought process.

Standard sections		actions	Thought process
Ph I	User information, Historical research, setting, Gov. records	Collect, review information develop a plan to perform the phase I	CSM 1
	Site reconnaissance Interviews	Inspect the site and interview knowledgeable individuals	CSM 2
	Phase I report - Findings opinions conclusions	Prepare report	CSM 3
Ph II	User needs/scope contractual considerations	Obtain info: plan Phase II prepare proposal/contract	CSM 4
	Developing Scope of Work	Refine scope/develop sampling plan, Conduct screening/sampling	CSM 5
	Assessment activities		CSM 6
	Evaluation of data	Review and analyze results	CSM 7
	Interpretation of results Phase II report	Present results/report	CSM 8

ESA scope of work that complies with the ASTM **E1903** Standard and meets the user's needs, the consultant must develop a thorough understanding of the user's needs.

Figure 11.2 depicts a conceptual representation of the relationship of the standards, the tasks, and the conceptual site model thinking process. The process builds as new information becomes available and leads to multiple reiterations of the conceptual site model. The conceptual site model begins with the Phase I assessment. As the assessment process progresses, the reiterations of the conceptual site model become more robust until a decision is reached to proceed with Phase II. The Phase II assessment may only use an applicable part of the conceptual site model developed in the Phase I portion of the process. As the Phase II process continues, the reiterations of the conceptual site model expand to include more information, once again creating a more robust model.

11.3   Objectives of **E1903-11**

Developing a Phase II ESA scope of services in conformance with the **E1903** Standard Practice is a fact-finding process in which the

FIG. 11.1   Potential issues that give rise to and define the scope of Phase II ESA.

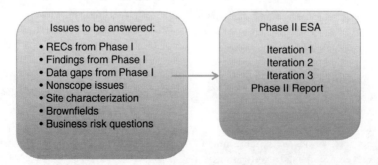

consultants seek out an understanding of users' needs and apply their knowledge, experience, and expertise in selecting a course of action that enables the users to meet their business objectives.

The 2011 revisions streamlined and simplified some of the decision-making process and created several choices with respect to possible objectives of the Phase II assessment. The six choices of objectives are outlined in **Fig. 11.3**.

Today, few industrial properties are found without some level of contamination. Although the innocent purchaser defense was a central issue during the development of the initial standards for the Phase I ESAs and Phase II ESAs, the standards have since evolved into useful business tools with multiple purposes. Both standards reflect historical development in the marketplace. By the time the first Phase II standard was written, the common use of the Phase I standard for business decisions was recognized. The Phase II Standard was first published in 1997, a full 4 years after the initial publication of the first edition of the **E1527**. E1527 vaguely discussed aspects of a business decision, and 4 years later **E1903** defined the term, making it an integral part of the standard. In the 2011 version, the definition of "business decision" was replaced with a definition of "business environmental risk." This is shown in **Fig. 11.4**.

The subsequent revisions of the Phase I standard also reflected some of the changing applications of the standards with respect to business environmental risk. The nature of the transaction, the risk tolerance of the purchaser, the purchase price, and many other factors influence the selection of the appropriate scoping strategy. Today, the consultant must be aware of the needs of the marketplace and structure the scope of Phase II ESAs in a manner that affords the client sufficient information to reach business decisions.

One of the purposes of **E1903** is to establish a process to provide the user with reliable information. The consultant will be the source of this information. Qualification, knowledge, and experience of the EP play a significant role in terms of reliability of the information. Inexperience can result in significant liability exposure for the consulting firm and the user. However, because of the market and pricing pressures, the EPs assigned to perform the Phase I and II ESAs work often comprise the junior, least experienced, and lowest cost staff in the firm. Senior staff members need to carefully review the manner in which the Phase I and II ESAs are being conducted to ensure compliance with standards and conformity with the firm's policies. Consultants that have been trained and have experience in the application of the ESA standards should understand significant differences and nuances in the ESA standards and provide appropriate direction to the junior staff.

As the industry matured through the late 1990s, many properties were identified with various types and various levels of contaminants. When it became obvious that cleaning up the contamination to "original" or to below-detection limits was impractical if not impossible, the risk-based corrective approach became more accepted at the federal and state level. With limited cleanup funding available, many properties were sitting in limbo awaiting further regulatory action. Some properties ended up with institutional or engineering controls. For these properties to be sold or redeveloped, the Phase II questions did not deal with whether the contamination was present, but rather how and whether the contaminant levels will affect the redevelopment of the subject property. This obviously dictated a different approach to the Phase II assessments than what would typically be associated with establishing applicable limited liability defenses. Many of these issues are addressed further in **Chapter 3** and illustrated with numerous case studies.

E1903 does not specify which methodologies should be used. Many different standards or methodologies may be applicable. This is the area where the users are looking for help from the

FIG. 11.3 List of Phase II ESA objectives.

1.2 Objectives—This practice is intended for use in any situation in which a user desires to obtain sound, scientifically valid data concerning actual property conditions. Without attempting to define all such situations, this practice specifically contemplates that users may seek such data to meet the following objectives:
1.2.1 Objective 1—Assess whether there has been a "release" of "hazardous substances" within the meaning of CERCLA for purposes including Landowner Liability Protections (i.e., Innocent Landowner, Bona Fide Prospective Purchaser, and Contiguous Property Owner).
1.2.2 Objective 2—Provide information relevant to identifying, defining, and implementing landowner "continuing obligations," or the criteria established under CERCLA (e.g., taking reasonable steps to prevent or limit exposures to previously released hazardous substances) for maintaining the CERCLA Landowner Liability Protections.
1.2.3 Objective 3—Develop threshold knowledge of the presence of substances on properties within the scope of the CERCLA definition of a "Brownfields site" and as required for qualifying for Brownfields remediation grants from the EPA Brownfields Program.
1.2.4 Objective 4—Provide information relevant to identifying, defining, and evaluating property conditions associated with target analytes that may pose risk to persons on the property.
1.2.5 Objective 5—Provide information relevant to evaluating and allocating business environmental risk in transactional and contractual contexts, including transferring, financing, and insuring properties and due diligence relating thereto.
1.2.6 Objective 6—Provide information to support disclosure of liabilities and contingent liabilities in financial statements and securities reporting.

FIG. 11.4 **E1903-11** definition of business environmental risk.

3.1.6 *Business environmental risk*: A risk which can have a material environmental or environmentally-driven impact on the business associated with the current or planned use of a parcel of commercial real estate, not necessarily limited to those environmental issues investigated in accordance with this practice.

consultants in determining what methodology will bring out the information the users are seeking. To choose the right method, the EPs must know the extent of users' objectives and possess sufficient training and experience to implement the ESA.

The objectives set during the scoping process should enable the consultants to draw conclusions about the contamination of the subject property. The conclusions could simply be that the site has contamination or that there is no reasonable basis to suspect that the site is contaminated. Alternatively, the conclusions could relate to the potential impacts on the expected future uses or redevelopment of the property. The conclusions should be such that they provide sufficient information to the user to make a business decision. For the conclusion to satisfy users' requirements, the consultant must fully understand users' needs before finalizing the scope of the Phase II ESA.

11.4 Users' Needs

As we pointed out, in many cases establishing limited liability protection defenses is not the objective of the user. E1903 recognizes that in many situations the extent of the assessment will be based on the business objectives of the user. The EPs must be cognizant of these objectives when establishing the scope of Phase II ESAs with the users. **Figure 11.5** shows key sections of the standard that summarize the concepts.

11.5 Limitations

None of the ESAs, including Phase II ESAs, can eliminate all uncertainty. This is why the Phase II ESA objectives are established on the degree of uncertainty acceptable to the user. For some clients, mere detection of contamination on the subject property may be sufficient to make a business decision. Under different circumstances, the extent of the contamination may be of importance in guiding the users' business judgments. Obviously, the scope of Phase II ESAs under each of the circumstances will vary greatly.

The risk tolerance varies from user to user. Some users are far more risk tolerant than others. The amount of risk tolerance will also affect the scope of a Phase II ESA. When multiple users are involved in the transaction, the consultant must identify the key user and tailor the scope of the Phase II ESA to meet those needs. For example, the purchaser may be considerably more risk tolerant than the lender. Although they are both users, the ESA must satisfy the needs and risk tolerance of the lender. Failure to do so will result in "no loan" and therefore "no sale."

A common misconception among both the users and EPs is that Phase II ESAs always involve sample collection. In actuality, Phase II ESAs may consist of conducting additional interviews or gathering additional historical information that can lead to a satisfactory answer to the question posed. For example, if an in-depth interview with a governmental official and a review of additional file materials reveals that the agency has reviewed the contamination on the subject property and has determined that no remedial action is required, then this information alone may be sufficient for the user to reach a decision.

Another example is when the consultant raises concerns with the uses of the neighboring property. This is demonstrated in the case study in **Fig. 11.6**.

In the case study in **Fig. 11.6**, any other result except the negative for contaminants will raise questions and present problems that this type of investigation was not designed to answer. With today's detection limits in parts per billion, and the ability to test for hundreds of potential contaminants, invariably the results almost never come out below detection. Once data come back, they have to be evaluated and interpreted. Are the data representative? Are they part of the background? Did the methodology affect the reliability of the result? That evaluation can be difficult if not impossible unless the potential for unexpected results is included in the sampling design. Let us consider some of the reasons why.

Any sampling approach has limitations. Those limitations and their effect on results should be discussed up front with the client.

In our example, taking three direct push samples in a line along a property boundary will not allow a good evaluation of background conditions of soil and groundwater, and it is unlikely to provide valuable information on groundwater flow. In addition, these types of "quick and dirty" Phase II investigations are often

---

FIG. 11.5 Section of **E1903** describing concept of users' needs.

> 1.4 *Needs of the User*—The *user* and *Phase II Assessor* must have a mutual understanding of the context in which the *Phase II ESA* is to be performed and the objectives to be met by the investigation, i.e., the specific questions to be answered or problems to be resolved by the *Phase II ESA*. The scope of Phase II activities must be defined in relation to those objectives.
> 1.4.1 The degree of confidence desired by the *user* influences the scope of the investigation and the evaluation of data. More extensive testing and more iterations of sampling and analysis may be needed if the objectives require detailed conclusions with high confidence. Less testing and fewer iterations of sampling and analysis may be needed if the objectives of the assessment include only general conclusions.

FIG. 11.6 Case study on expected results.

> Phase I report identified a transfer station for waste management on adjacent property, and the consultant raised a concern as to how that use may impact the subject property. The consultant recommended a seemingly simple approach to take three direct push samples of soil and groundwater along the boundary of the property. However, this has invariably shown to be an inappropriate approach. In many instances, the parties suggest this type of approach when they expect a certain result. The consultant may have suggested the simple approach, expecting that the sampling will come back negative for contaminants, and thus will be a hard data for the client to show that there is no contamination. That was the expected result.

susceptible to shortcuts in obtaining soil and groundwater parameters and often include quality control sampling shortcuts that can cause interference due to parameters such as pH, turbidity, contamination, and cross-contamination. These are factors that can make evaluation and interpretation of the sample results almost impossible.

When the results are potentially damaging to the real estate transaction, the typical comeback from the consultant is that they need more money to get more, and better, data. At this point, the results are causing problems, and a host of issues including litigation can surface. Before making a decision to get some "quick" data, a sound scientific approach should be established. This may involve interviewing the operator, a visit to the neighboring transfer station facility, and reviewing regulatory data with respect to the neighboring facility. This may prove to be sufficient information to resolve the user's business risk concerns without generating test results that lack sufficient information to be accurately interpreted.

The **E1903** Standard Guide did not require a written report. The 2011 revision of **E1093** now requires a report. This can create issues, particularly if the information and data generated as part of the process cause the user to walk away from the property. At that point, he has no need for a report and may terminate the relationship and refuse to pay for report preparation. Although many scope issues and results of the investigation may be communicated verbally, the consultant should adequately document communications, agreement on the scope issues, methods used, and results obtained.

The consultants' professional code of ethics as well as local, state, and federal law may also impose reporting obligations on the EPs that need to be communicated to the user early in the scoping process. The **E1903-11** clearly states that it is not intended to supersede regulatory requirements, as shown in **Fig. 11.7**.

11.6 Summary

In this chapter, we learned that **E1903** changed in 2011 from a standard guide to a standard practice. This represents a significant

FIG. 11.7 **E1903-11** scope limitations.

> 1.5 *Limitations*—This practice is not intended to supersede applicable requirements imposed by regulatory authorities. This practice does not attempt to define a legal standard of care either for the performance of professional services with respect to matters within its scope, or for the performance of any individual *Phase II Environmental Site Assessment*.

change to the way in which the ESAs are conducted. The scoping process enables the consultant and the user to establish a mutually acceptable process that is based on scientific principles. The consultant designs the process. A thorough understanding of the objectives, needs, limitations, and requirements is crucial in the determination of the appropriate scope of the work. In this chapter, we introduced the potential diversity of Phase II ESAs, which will be further demonstrated by examples and case studies in the subsequent chapters. The issues in this chapter are closely related to the topics also addressed in **Chapter 3** on the application and uses of **E1903**.

References

[1] ASTM Standard **E1527**, Standard Practice for Environmental Site Assessments: Phase I Environmental Site Assessment Process, *Annual Book of ASTM Standards,* ASTM International, West Conshohocken, PA, 2013.

[2] ASTM Standard **E1903**, Standard Practice for Environmental Site Assessments: Phase II Environmental Site Assessment Process, *Annual Book of ASTM Standards,* ASTM International, West Conshohocken, PA, 2011.

[3] ASTM Standard **E1903**, Standard Guide for Environmental Site Assessments: Phase II Environmental Site Assessment Process, Historical Standard, *Annual Book of ASTM Standards,* ASTM International, West Conshohocken, PA, 1997.

Chapter 12 | Expanding the Environmental Vocabulary

12.1 Standardized Terminology of E1903

In Part 1 of the manual, we introduced and explained the importance of standardized terminology. The **E1903** Standard [1] has 54 standard definitions—17 more than the previous 2002 standard guide. During the conversion of the guide to a standard, 16 terms were not carried over and 32 new terms were defined.

Of the 16 deleted terms, various simple terms such as "aquifer," "cuttings," "disposal," and "sample medium" were not carried over because they were not needed within the framework of the scientific method. Many of the new terms did relate to the scientific method framework (e.g., "conceptual model," "exposure point," "likely release area," and "migration pathway"). Other new terms defined in the Phase II standard show the continued link to CERCLA (i.e., the Comprehensive Environmental Response, Compensation, and Liability Act), including "All Appropriate Inquiry," "Bona Fide Prospective Purchaser," "Contiguous Property Owner," "continuing obligations," "Innocent Landowner Defense," and "Landowner Liability Protections"). A significant new term is the addition of the definition of a "Phase II assessor."

Remaining, unchanged terms and definitions are related to sampling and site investigation activities that may be associated with Phase II Environmental Site Assessments (ESAs).

Closer examination of some of the definitions in the **E1903** Standard provides additional insights into the intent and scope of the standard practice. In **Chapter 1**, we already reviewed the definition of a business environmental risk. The resolve of the ASTM E50 Committee to include it in the **E1903** Standard demonstrates that business risk management considerations are an integral part of the Phase II ESA process.

Acronyms such as the U.S. Environmental Protection Agency (EPA) All Appropriate Inquiry Final Rule (AAI) are defined in the Terminology section of **E1903**. This term is used in CERCLA and interpreted by the EPA AAI, 40 CFR Part 312, and it clearly shows the relationship of the Phase II ESA standard with CERCLA. It also indicates how federal or state cleanup regulations and requirements may become part of a specific Phase II ESA process.

"Chain of custody" and "quality assurance/quality control" (QA/QC) and similar terms defined in **E1903** point to the fact that Phase II ESAs generally include sampling of environmental media on the subject site.

12.2 Key New Terms to the 2011 E1903 Standard Practice

Perhaps the most significant change from a standard guide to the standard practice was the restructuring basis of a Phase II ESA around the scientific method, which is generally understood as

- What is the question (i.e., objective)?
- Research (Phase I or similar effort)
- Hypothesis (What do you expect from the data?)
- Experimental strategy (scope of work)
- Sample/test (implement site assessment plan)
- Analyze results (QA/QC; Did you answer the questions?)
- Conclusion (The answer is X. Now what?)
- Report (i.e., Phase II)

The **E1903** definition of the scientific method as it appears in the standard is depicted in **Fig. 12.1**.

The framework within which a hypothesis is developed is the conceptual model. Conceptual models can range from simple statements such as "Are there solvents in the soil around the dry-cleaning machine?" to detailed flow charts that define sources, receptors, and open and closed pathways. ASTM has several standards relating to conceptual models, including **E1689** *Guide for Developing Site Models for Contaminated Sites* and **E2531** *Guide for Development of Conceptual Site Models and Remediation Strategies*

FIG. 12.1 Scientific method definition from the **E1903** Standard Practice.

3.1.51 *The scientific method*—principles and procedures for systematic discovery, which involve recognizing and stating a problem for which a solution is sought, formulating an hypothesis that might resolve the problem (which hypothesis is consistent with the body of knowledge available), collecting objective and reproducible data by performing an investigation to test the hypothesis, and interpreting the data to validate or refute the hypothesis; and, if the hypothesis cannot be validated, revising the hypothesis consistent with the updated body of knowledge and conducting an iteration of the procedure.

FIG. 12.2   Conceptual model definition from the **E1903** Standard Practice.

> 3.1.11 *Conceptual model*—a representation of hypothesized current site conditions, which describes the physical setting characteristics of a site and the likely distribution of *target analytes* that might have resulted from a known or likely *release*, and which is based on all *reasonably ascertainable* information relevant to the objectives of the investigation and the professional judgment of the *Phase II assessor*.

FIG. 12.3   Likely release area definition from the **E1903** Standard Practice.

> 3.1.28 *Likely release area*—a place where a *Phase II assessor* judges it likely that *target analytes* were first introduced into *environmental media* as a result of a *release* such that the *target analytes* may now be *present* in *environmental media* at the *property*. *Likely release areas* can include, but need not be limited to, *recognized environmental conditions* identified in a *Phase I ESA* conducted in accordance with Practice **E1527** or Practice **E2247**.

for Light Non-Aqueous Phase Liquids Released to the Subsurface. The **E1903** task group created a new definition for a conceptual model that is specific to the Phase II site assessment. As is depicted in **Fig. 12.2**, the conceptual model drives the scope of work.

Another newly defined term in the **E1903** is "likely release area." The likely release area also drives the scope of work. An understanding of the place where potential contamination was first released to the environment will help in the development of the Phase II ESA sampling methodology based on the media affected (e.g., soil around a pump island, pond water adjacent to slag heap, air downwind from a cement stack, sediments in a bay near an industrial outfall) and the actual sampling depths in the case of soil or water. The definition from the standard is depicted in **Fig. 12.3**. An important aspect of the definition is that likely release areas can include, but need not be limited to, recognized environmental conditions (RECs) identified in a Phase I ESA. This demonstrates that the Phase I and II scopes may not concentrate only on RECs.

12.3   Environmental Professional and Phase II Assessor Definitions and Special Qualifications Requirements

E1527 [2] and **E1903** contain definitions of the people who complete Phase I and II ESAs, environmental professionals (EPs) and Phase II assessors, respectively. In the 2005 version of the **E1527** standard, the definition of the EP changed significantly from the previous versions. There is a continued link in the definitions because the root of the Phase II assessor definitions is founded on the **E1527** definition of an EP. The Phase II assessor is an expansion of an EP that can be in responsible charge to devise, plan, implement, and interpret Phase II ESAs. The Brownfields Amendments drove the changes. A new definition of a Phase II assessor was developed that establishes basic areas of understanding to complete a Phase II ESA (i.e., scientific method, hydrogeology, geochemistry, environmental investigation/exploration techniques, chemical testing data, commercial and industrial operations). The definitions continue to reflect the trend in the industry of increasing the professional requirements when going from Phase I to Phase II assessments. The Phase II EP is likely to have sufficient qualification to meet the requirements of the Phase I EP, but not all Phase I EPs will meet the requirements to practice as the Phase II professionals. Because of the potentially significant and frequent revisions of the standards, it is imperative that the professionals using the standards stay current on the applicable changes. Some jurisdictions have licensing requirements for individuals who perform certain environmental investigation activities included in a Phase II ESA.

The current differences and relationship in the definitions between the two standards can be seen in **Fig. 12.4**.

In the Phase I ESA, the EP has to meet the requirements of the federal regulation. The Phase II assessor definition increases the responsibility and areas of understanding. It requires the EPs to be able to prepare and implement the Phase II ESA. The **E1903-11** definition also includes discussion language with respect to potential licensing requirements. This is generally related to federal or state regulated sampling or investigation activities that may end up being included in the Phase II ESA. For example, a Phase II ESA

FIG. 12.4   Definitions of EP and Phase II assessor from the **E1527** and **E1903** Standards.

> **3.3.29 Environmental professional:** A person meeting the education, training, and experience requirements as set forth in 40 CFR 312.10(b). See Appendix X2. The person may be an independent contractor or an employee of the user.
>
> **3.1.33 Phase II assessor:** A person meeting the definition of an *environmental professional* as provided in Section 3.2.29 of Practice **E1527** and possessing sufficient education, professional training, and relevant experience to conduct or be in responsible charge of environmental investigations and other activities in accordance with this practice, and to interpret the resulting data to develop opinions and conclusions regarding the *presence* of *target analytes* in *environmental media* in connection with the *property* in question. An individual's status as a *Phase II assessor* may be limited to the type of assessment to be performed. Overall, a *Phase II assessor* should understand and be experienced in pertinent aspects of *the scientific method*, hydrogeology, geochemistry, environmental investigation/exploration techniques, interpretation of *chemical testing* data, and commercial and industrial operations pertaining to the use and handling of site-specific *target analytes* and production and handling of associated wastes. The *Phase II assessor* may be an independent contractor or an employee of the *user*. Some jurisdictions may have licensing requirements for individuals who perform certain environmental investigation activities included in a *Phase II ESA*.

FIG. 12.5 Excerpts from EPA evaluation report.

> The EPA Office of the Inspector General (OIG) Evaluation Report No. 11-P-0107—*EPA Must Implement Controls to Ensure Proper Investigations Are Conducted at Brownfields Sites*—issued February 14, 2011 states that based on a review of "35 AAI [All Appropriate Inquiry] reports we reviewed, from three EPA regions, none contained all the required documentation elements…. Improper AAI investigations introduce risk that the environmental conditions of a property have not been properly or adequately assessed, which may lead to improper decisions about appropriate uses of Brownfields properties. Ultimately, threats to human health and the environment could go unrecognized…If conditions merit, EPA is authorized to take back funds from noncompliant grantees."

under the EPA's oversight requires the professional to have specific qualifications for certain portions of the assessment. Some states such as Connecticut, New Jersey, Nevada, and Massachusetts have credential programs (e.g., licensed site remediation professional [LSRP], certified environmental manager [CEM]). Many Brownfields programs also include special requirements.

EPA created the checklist of the specific Brownfields requirements because of its findings at Brownfields sites in the Office of Inspector General Evaluation Report No. 11-P-0107. Excerpts from the report are depicted in **Fig. 12.5**. This noncompliance was largely related to terminology and resulted in the checklist in **Fig. 12.6**.

Figure 12.6 is the EPA Reporting Requirement Checklist associated with the completion of Phase I ESAs that is an example of specific requirements.

Many of the nonscope issues of Phase I assessments may be included in the Phase II assessments, and special regulations will apply. Several states regulate asbestos and radon testing in public buildings. The State of Texas also regulates mold. In many states, integrity testing and evaluation of underground storage tanks can only be performed by state-certified personnel.

The requirement of being able to develop sound opinions in Phase II ESAs emphasizes the focus on sound scientific supporting data when rendering opinions in Phase II ESAs.

12.4 Referenced Terminology

The Phase II standard references the three related ASTM standards: **E1527** Practice for Environmental Site Assessments: Phase I Environmental Site Assessment Process; **E1528** Practice for Limited Environmental Due Diligence: Transaction Screen Process; and **E2247** Practice for Environmental Site Assessments: Phase I Environmental Site Assessment Process for Forestland or Rural Property. Another useful source of terminology is ASTM **D653** *Terminology Relating to Soil, Rock, and Contained Fluids* [**3**]. This standard contains definitions that were prepared jointly by ASTM and the American Society of Civil Engineers. The **D653** Terminology Standard is a 52-page document that includes an alphabetical listing of terms relating to soil, rock, and contained fluids. It is under the jurisdiction of ASTM Committee D-18 on soil and rock. An understanding of the terms is often crucial to the required experience in hydrogeology, geochemistry, and environmental investigations or exploration.

E1903 and **D653** are included in a compilation of Standards entitled *ASTM Standards on Environmental Site Assessment Process*. [**4**]. This is an electronically available compilation of standards, and the standards contain terminology sections further expanding the Phase II assessor's environmental vocabulary. The compilation is an excellent reference publication for Phase II assessors involved in Phase II ESAs. The subscription for this electronically available compilation of standards provides the user of the standards with the latest version of the standards and makes standards searching an easy task.

A considerable amount of environmental terminology also comes from non-ASTM sources. Some of these sources of non-ASTM environmental terminology are listed in **Table 12.1**. The table is from ASTM **D5730** *Standard Guide for Site Characterization for Environmental Purposes with Emphasis on Soil, Rock, the Vadose Zone, and Groundwater*[**5**]. This standard was withdrawn in 2013 because the task group did not update it within the required cycle of 8 years. Even as a historical standard, it can be a useful source of various assessment and investigation methodologies and non-ASTM terminology related to the field of environmental science.

Terminology of **E1527** Phase I ESA is incorporated in the Phase II Standard by reference. Many of the terms introduced in **E1527** are used in **E1903**.

12.5 Summary

In this chapter, we discussed the standardized terminology defined in **E1903**. We learned that the Phase II ESAs may involve numerous issues regulated by other standards or by federal, state, or local regulations. We identified numerous sources of related environmental terminology and introduced some of the assessment standards specifically referenced in **E1903**.

References

[1] ASTM Standard **E1903**, Standard Practice for Environmental Site Assessments: Phase II Environmental Site Assessment Process, *Annual Book of ASTM Standards*, ASTM International, West Conshohocken, PA, 2011.

[2] ASTM Standard **E1527**, Standard Practice for Environmental Site Assessments: Phase I Environmental Site Assessment Process, *Annual Book of ASTM Standards*, ASTM International, West Conshohocken, PA, 2013.

[3] ASTM Standard **D653** Terminology Relating to Soil, Rock, and Contained Fluids, *Annual Book of ASTM Standards*, ASTM International, West Conshohocken, PA, 2011.

[4] ASTM Standards on Environmental Site Assessment Process, ASTM International, West Conshohocken, PA, 2013.

[5] ASTM Standard **D5730**, Standard Guide to Site Characterization for Environmental Purposes with Emphasis on Soil, Rock, the Vadose Zone, and Groundwater, *Annual Book of ASTM Standards*, ASTM International, West Conshohocken, PA, 2004, withdrawn 2013.

FIG. 12.6 EPA requirement and checklist.

All Appropriate Inquiries Rule:
Reporting Requirements Checklist for Assessment Grant Recipients

WHAT IS "ALL APPROPRIATE INQUIRIES"?

"All appropriate inquiries," or AAI, is a process of evaluating a property's environmental conditions and assessing the likelihood of any contamination. Every Phase I assessment conducted with EPA Brownfields Assessment Grant funds *must* be conducted in compliance with the All Appropriate Inquiries Final Rule at 40 CFR Part 312. The All Appropriate Inquiries Final Rule provides that the ASTM E1527-05 standard is consistent with the requirements of the final rule and may be used to comply with the provisions of the rule. In addition, all appropriate inquiries must be conducted in compliance with the Final Rule or the ASTM standard to obtain certain protections from liability under the federal Comprehensive Environmental Response, Compensation and Liability Act (CERCLA, aka Superfund).

WHAT ARE THE DOCUMENTATION REQUIREMENTS FOR ALL APPROPRIATE INQUIRIES?

The final rule requires that the results of an AAI investigation be documented in a written report. The specific reporting requirements for all appropriate inquiries are provided in 40 CFR §312.21 (*Results of Inquiry by an Environmental Professional*) and §312.31 of the final rule and in §12 of ASTM E1527-05. All AAI final reports *must*, at a minimum include the following:

1. An *opinion* as to whether the inquiry identified conditions indicative of releases or threatened releases of hazardous substances, and as applicable, pollutants and contaminants, petroleum or petroleum products, or controlled substances, on, at, in, or to the subject property.

2. An identification of *"significant" data gaps* (as defined in §312.10 of AAI final rule and §12.7 of ASTM E1527-05), if any, in the information collected for the inquiry. Significant data gaps include missing or unattainable information that affects the ability of the environmental professional to identify conditions indicative of releases or threatened releases of hazardous substances, and as applicable, pollutants and contaminants, petroleum or petroleum products, or controlled substances, on, at, in, or to the subject property. The documentation of significant data gaps must include information regarding the significance of these data gaps.

3. *Qualifications* and *signature* of the environmental professional(s). The environmental professional must place the following statements in the document and sign the document:

    - *"[I, We] declare that, to the best of [my, our] professional knowledge and belief, [I, we] meet the definition of Environmental Professional as defined in §312.10 of this part."*

    - *"[I, We] have the specific qualifications based on education, training, and experience to assess a property of the nature, history, and setting of the subject property. [I, We] have developed and performed the all appropriate inquiries in conformance with the standards and practices set forth in 40 CFR Part 312."*
    Note: Please use either "I" or "We."

4. In compliance with §312.31(b) of the AAI final rule and §12.6.1 of ASTM E1527-05, the environmental professional must include in the final report an *opinion regarding additional appropriate investigation*, if the environmental professional has such an opinion.

For each Phase I Environmental Site Assessment, or AAI, conducted using EPA Brownfields Grant monies, the grantee must complete, sign, date, and submit the following check list with any AAI reports submitted to EPA.

Also see U.S. EPA's website at www.epa.gov/brownfields for additional information.

1

FIG. 12.6 *(Continued).*

Checklist for Phase I Site Assessments Conducted using EPA Brownfields Assessment Grant Funds

Contact Information

Grantee Name: _____

Grant Number: _____

ACRES Property ID: _____

Program Manager Name: _____
(Point of Contact)

Contact Phone Number: _____

Name / Address of Property Assessed: _____

Checklist

Please indicate that each of the following All Appropriate Inquiries documentation requirements were met for the Phase I assessment conducted at the above listed property:

☐ An *opinion* as to whether the inquiry has identified conditions indicative of releases or threatened releases of hazardous substances, and as applicable, pollutants and contaminants, petroleum or petroleum products, or controlled substances, on, at, in, or to the subject property.

☐ An identification of *"significant" data gaps* (as defined in §312.10 of AAI final rule and §12.7 of ASTM E1527-05), if any, in the information collected for the inquiry, as well as comments regarding the significance of these data gaps. Significant data gaps including missing and unattainable information that affects the ability of the environmental professional to identify conditions indicative of releases or threatened releases of hazardous substances, and as applicable, pollutants and contaminants, petroleum or petroleum products, or controlled substances, on, at, in, or to the subject property.

☐ *Qualifications* and *signature* of the environmental professional(s). The environmental professional must place the following statements in the document and sign the document:
  ☐ "[I, We] declare that, to the best of [my, our] professional knowledge and belief, [I, we] meet the definition of Environmental Professional as defined in §312.10 of this part."

  ☐ "[I, We] have the specific qualifications based on education, training, and experience to assess a property of the nature, history, and setting of the subject property. [I, We] have developed and performed the all appropriate inquiries in conformance with the standards and practices set forth in 40 CFR Part 312."
  Note: Please use either "I" or "We."

☐ In compliance with §312.31(b) of the AAI final rule and §12.6.1 of ASTM E1527-05, the environmental professional must include in the final report an *opinion regarding additional appropriate investigation*, if the environmental professional has such an opinion.

_____   _____
Signature of Grantee Program Manager   Date

Brownfields Fact Sheet *Solid Waste* *EPA 560-R-11-030*
AAI: Reporting Requirements and *and Emergency* *June 2011*
Grantee Checklist *Response (5105)* *www.epa.gov/brownfields*

TABLE 12.1 Major Non-ASTM References on Terminology Related to Environmental Site Characterization from the Appendix of **E5730**

X3. MAJOR NON-ASTM REFERENCES ON TERMINOLOGY RELATED TO ENVIRONMENTAL SITE CHARACTERIZATION

Allaby, A. and Allaby, M., *The Concise Oxford Dictionary of Earth Sciences*. Oxford University Press, Oxford, UK, 1990.

American Society of Agricultural Engineers, *Glossary of Soil and Water Terms*. American Society of Agricultural Engineers, St. Joseph, MI, 1967.

Bates, R. and Jackson, J. (eds), *Dictionary of Geological Terms*, 3rd ed. AGI, Washington, DC, (Supersedes Weller (1960)), 1984.

Castany, G. and Margat, J., *Dictionnaire Franais D'Hydrogéolgie*, BRGM, Orléans, 1977.

Interagency Advisory Committee, *Subsurface-Water Flow and Solute Transport, Glossary of Selected Terms*. Draft report prepared by Subsurface-Water Glossary Working Group, Ground Water Subcommittee, 1988.

International Society for Rock Mechanics, *Final Document on Terminology, English Versions*, Committee on Terminology, Symbols and Graphic Representation, 1972.

Laney, R. L. and Davidson, C. B., *Aquifer Nomenclature Guidelines*, U.S. Geological Survey Open File Report 86-534, 1986.

Langbein, W. B. and Iseri, K. T., *General Introduction and Hydrologic Definitions*. U.S. Geological Survey Water Supply Paper 1541-A, 1960.

Lo, S. S., *Glossary of Hydrology*, Water Resources Publications, Highland Ranch, CO, 1992.

Lohman, S. W., et al, *Definitions of Selected Ground-Water Terms—Revisions and Conceptual Refinements*, U.S. Geological Survey Water-Supply Paper 1988, 1972.

Meinzer, O. E., *Outline of Ground Water Hydrology with Definitions*, U.S. Geological Survey Water Supply Paper 494, 1923.

Michel, J.-P. and Fairbridge, R. W., *Dictionary of Earth Sciences*, John Wiley & Sons, New York, 1992.

Moore, W. G., *A Dictionary of Geography*, 4th edition, Penguin Books, Baltimore, MD, 1968.

National Geodetic Survey, *Geodetic Glossary*, U.S. Department of Commerce, 1986.

Parker, S. P. (ed), *Dictionary of Scientific and Technical Terms*, 4th ed., McGraw-Hill, New York, 1989.

Pfannkuch, H. O., *Elsevier's Dictionary of Hydrogeology*, Elsevier, NY, 1969.

Poland, J. F., et al, *Glossary of Selected Terms Useful in Studies of the Mechanics of Aquifer Systems and Land Subsidence Due to Fluid Withdrawal*, U.S. Geological Survey Water Supply Paper 2025, 1972.

Porteous, A., *Dictionary of Environmental Science and Technology*, revised edition, John Wiley & Sons, New York, 1992.

Soil Conservation Service (SCS), *Glossary of Selected Geologic and Geomorphic Terms*, U.S. Department of Agriculture, Soil Conservation Service Western Technical Service Center, Portland, OR, 1977.

Soil Science Society of America, *Glossary of Soil Science Terms*, SSSA, Madison, WI, 1987.

Stevenson, L. H., and Wynen, B., *The Facts on File Dictionary of Environmental Sciences*. Facts on File, New York, NY, 1991.

Titelbaum, O. A., *Glossary of Water Resources Terms*, Federal Water Pollution Control Administration, 1970.

UNESCO, *International Glossary of Hydrology*, WMO/OMM/BMO No. 385, 1974.

U.S. Environmental Protection Agency (EPA), *Draft Glossary of Quality Assurance Related Terms*, Office of Research and Development, Sept. 29, 1988.

U.S. Geological Survey, *Federal Glossary of Selected Terms: Subsurface-Water Flow and Solute Transport*, Office of Water Data Coordination, USGS, Reston, VA, 1989.

Whitten, D. G. A. and Brooks, J. R. V., *The Penguin Dictionary of Geology*, Penguin Books, Baltimore, MD, 1972.

Weller, J. M. (ed), *Glossary of Geology and Related Sciences with Supplement*, 2nd edition. American Geological Institute, Washington, DC, Supplement 72 pp. (Superseded by AGI (1984)), 1960.

Chapter 13 | Application and Uses of E1903

13.1 Intended Uses and Applications of E1903 Uses

The 2011 revision of the Phase II standard significantly expanded the intended uses of the standard. The E1903 Standard Practice [1] is still a voluntary standard, meaning that unless there is a state or federal program that mandates the use of the standard, there is no requirement that the consultants follow this standard. It is intended to be used by parties that wish to evaluate known releases or likely release areas. The practice was developed to meet the need for a written, practical reference describing a scientifically sound approach that can be consistently applied for conducting a Phase II Environmental Site Assessment (ESA).

The original objective of E1903 was to provide guidance for building upon the recognized environmental conditions (RECs) typically identified in either the Phase I ESA or through some other avenues, and that can still be a common desired objective. In **Chapter 11**, we identified the six primary objectives that typically drive the application of the E1903 Standard. They are listed in **Table 13.1**.

E1903 does not define the scope of work, the number of borings, or the analytical methods. The value of transitioning E1903 from a standard guide to a standard practice was to provide a consistent framework for conducting a Phase II ESA for any given objective. It specifies procedures based on the "scientific method." **Figure 13.1** is a simple depiction of the scientific method approach.

13.2 Applicability of E1903-11

The scope of a Phase II ESA is specific to the site and context. The assessment process defined by this practice is intended to generate sound, objective, and defensible information sufficient to satisfy diverse user objectives. The following series of case studies provides examples that demonstrate the variety of possibly scenarios that would be appropriate applications of E1903-11. The case studies are divided into Comprehensive Environmental Response, Compensation, and Liability Act (CERCLA) and non-CERCLA issue discussions.

13.2.1 CERCLA-RELATED CASE STUDIES

In the case study in **Fig. 13.2**, a Phase I ESA identified that the property was located in an area that was under investigation by the U.S. Environmental Protection Agency (EPA). The Phase II ESA was conducted on behalf of a potential buyer for the specific purpose of demonstrating that, through a reasonable level of inquiry, there was "no reason to know" that the CERCLA hazardous substances under conditions indicative of a release or threatened release were present.

Although the Innocent Landowner Defense (ILD) was a central issue during the development of the initial Phase I ESA and Phase II ESA standards, the 2002 Brownfields Amendments to CERCLA broadened the applicability of the Phase I ESAs and Phase II ESAs with the introduction of the Bona Fide Prospective Purchaser (BFPP). As discussed in Part 1 of this manual, the BFPP is a landowner liability protection (LLP) that was introduced with the 2002 legislation that allowed a purchaser of commercial real estate to acquire property with knowledge of contamination and to not be held liable for EPA response costs so long as certain requirements were met. One of those requirements is conducting the Phase I ESA or EPA All Appropriate Inquiries Final Rule (AAI) before acquiring the property. After acquisition, maintaining that liability protection requires adherence to "continuing obligations." LLPs remain available so long as the landowner

- Takes reasonable steps to stop/prevent any ongoing release,
- Provides full cooperation in connection with response actions,
- Is compliant with any land use restrictions or request for information,
- Does not impede the integrity of any institutional control used for response actions, and
- Provides legally required notices with respect to the discovery of hazardous substances.

In **Fig. 13.3**, the case study provides an example of how the Phase II ESA can be utilized to assist a user in meeting the requirements necessary to achieve and maintain a BFPP LLP.

In the case study, a Phase I ESA has identified that the property was located in an area known to be impacted by a National Priorities List (NPL) site. The Phase II ESA was conducted on behalf of a potential buyer for the specific purpose of evaluating where contaminants were currently located relative to planned construction activities and to assist in establishing any continuing obligations expected by the agency to maintain the LLP.

The 2002 Brownfields Amendments also provided a funding mechanism to assess and clean up Brownfields sites. The

TABLE 13.1 Phase II ESA Objectives

Objective 1—Assess whether there has been a "release" of "hazardous substances" within the meaning of CERCLA, for purposes including Landowner Liability Protections (i.e., Innocent Landowner, Bona Fide Prospective Purchaser, and Contiguous Property Owner).

Objective 2—Provide information relevant to identifying, defining, and implementing landowner "continuing obligations" or the criteria established under CERCLA (e.g., taking reasonable steps to prevent or limit exposures to previously released hazardous substances) for maintaining the CERCLA Landowner Liability Protections.

Objective 3—Develop threshold knowledge of the presence of substances on properties within the scope of the CERCLA definition of a "Brownfields site" and as required for qualifying for Brownfields remediation grants from the EPA Brownfields program.

Objective 4—Provide information relevant to identifying, defining, and evaluating property conditions associated with target analytes that may pose risk to persons on the property.

Objective 5—Provide information relevant to evaluating and allocating business environmental risk in transactional and contractual contexts, including transferring, financing, and insuring properties and due diligence relating thereto.

Objective 6—Provide information to support disclosure of liabilities and contingent liabilities in financial statements and securities reporting.

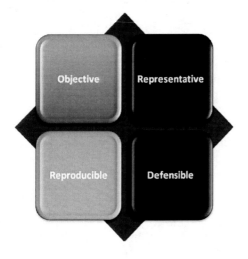

FIG. 13.1 Scientific method.

Brownfields Amendments define Brownfields sites as properties contaminated or potentially contaminated with substances not included in the definition of "hazardous substance" in CERCLA. Brownfields sites include properties contaminated with (or potentially contaminated with) hazardous substances, petroleum and petroleum products, controlled substances, pollutants, and contaminants. In the case of persons receiving federal Brownfields grant monies to conduct site assessment and characterization activities at Brownfields sites, the scope of the Phase I ESA/AAI may include these other substances in cases in which the terms and conditions of a specific grant or the cooperative agreement with a grantee designate a broader scope of investigation beyond CERCLA hazardous substances. EPA has stipulated that cleanup grant applicants must have a Phase II ESA report meeting the ASTM **E1903-11** Standard Practice, and the Phase II

FIG. 13.2 Classic CERCLA-related ILD Phase II ESA case study.

> **Phase II case study in which CERCLA LLPs are primary issue**
>
> A 5,000-acre area has been designated as a Comprehensive Environmental Response, Compensation, and Liability Information System (CERCLIS)-listed site because of potential impacts from historical uses as an arms manufacturing facility during World War II. The EPA is proceeding with preliminary assessment activities as part of the scoring process for determining if the site will meet the requirements for NPL status. The 5,000-acre area included a "buffer zone" where no arms manufacturing activities took place.
>
> A potential buyer is seeking to acquire several parcels of raw land for an anticipated industrial park development. The Phase I ESA discovered that the parcels are included within the boundaries of the 5,000-acre area CERCLIS-listed site. Although no known arms manufacturing activities were known to have taken place on or adjoining the parcels of interest, agency file documentation stated that there was potential that all dumping areas had not yet been identified.
>
> The property purchaser wanted to ensure that (1) the parcels of interest were not affected by the arms manufacturing activities and (2) EPA would not seek cost recovery in the event the site was ultimately designated NPL status. In cooperation with legal counsel, the Phase II assessor developed a scope of work that included soil and groundwater sampling to evaluate the presence of analytes consistent with EPA's investigation activities, and a series of shallow test holes to evaluate the potential for buried wastes. At the conclusion of the Phase II ESA activities, no constituents of concerns were identified on the subject property and no buried wastes were encountered.

assessment report must be completed before a grant proposal submission [2].

In the **Fig. 13.4** case study, prior assessment activities had been conducted that were appropriate for an industrial setting and appropriate for the risk tolerance of the parties involved. However, a clear understanding of the Brownfields grant program requirements is critical in defining an appropriate Phase II scope of work.

13.2.2 NON-CERCLA-RELATED CASE STUDIES

There are many instances when a Phase II ESA may be desired although CERCLA LLPs are not a concern. By the time the first Phase II standard was written, the common use of the Phase I standard for business decisions was recognized. The Phase II standard was first published in 1997, a full 4 years after the initial publication of the first edition of **E1527**. E1527 vaguely discussed aspects of a business decision, and 4 years later **E1903** defined the term, making it an integral part of the standard. In the 2011 version, the definition of business decision was replaced with a definition of "business environmental risk." This is shown in **Fig. 13.5**.

The subsequent revisions of the Phase I standard also reflected some of the changing applications of the standards with respect to

FIG. 13.3   Brownfields-related case study.

> **Phase II case study for new CERCLA BFPP LLP**
>
> A national retail store chain seeks to purchase property to develop a new retail facility. The Phase I ESA identifies historical uses of the property as agricultural land used for growing a wide variety of crops. No RECs were identified associated with past uses of the property; however, the adjoining property was identified as an NPL site, groundwater had been affected, and the groundwater plume has migrated onto a portion of the property of interest.
>
> At the time the retail store was considering its purchase of the property being affected by this NPL site, the EPA had concluded investigation activities, had installed a series of monitoring wells, and was evaluating remediation options. Because timing needs were critical, the retail chain representatives, the Phase II assessor, and legal counsel entered into discussions with the agency to formulate an appropriate scope of work that the Phase II assessor would implement to (1) fill any data gaps associated with the proposed new use of the property as a retail store and (2) determine soil and groundwater conditions in areas where excavation activities were expected to encounter impacted groundwater or influence groundwater flow.
>
> The buyer, the seller, EPA, and legal counsel provided input into the scope of work. Sample locations and analytes were specific to the known release from the NPL site, but they were also broadened to include an evaluation for potential residual agricultural chemicals (although the application of herbicides and pesticides in accordance with manufacturer directions is not considered a "release"; therefore, it would not meet the definition of a REC).
>
> At the conclusion of the Phase II ESA, the Phase II assessor found that groundwater was impacted at concentrations below established risk-based cleanup criteria for industrial property use but above concentrations appropriate for storm water discharge. Any de-watering activities would require treatment before discharge, and de-watering activities were to be carefully monitored to ensure minimal impact on groundwater flow. EPA agreed to relocate monitoring wells located in the proposed building footprint.
>
> EPA also entered into negotiations with the retail chain to negotiate the appropriate amount for the Windfall Lien (Windfall Liens are discussed in Part 1 of this manual).

FIG. 13.4   Case study related to Brownfields grantee issues.

> **Phase II case study for Brownfields grantee**
>
> A community was interested in converting an abandoned industrial building into artist lofts and affordable housing. At least two prior Phase I ESAs and one Phase II ESA had been conducted on the property for the purpose of financing ongoing commercial and light industrial operation. The prior Phase I ESAs had identified that the building was originally constructed as a paint manufacturer. A Phase II ESA had identified impacted soil and groundwater, but at concentrations that were acceptable for commercial use.
>
> The organization seeking to acquire the property was provided with the previous Phase I and Phase II ESAs; however, consistent with the needs and intended uses of past owners and operators, the ESA activities were focused only on hazardous substances and petroleum products that had been released to the environment. A scope of work was defined for the Brownfields grantee that included soil and groundwater sampling to evaluate if current conditions were appropriate for residential use, but it was also expanded to include evaluation of building materials such as lead-based paint, polychlorinated biphenyl (PCB)-containing paint, and asbestos-containing building materials. At the conclusion of the Phase II ESA, soil and groundwater impacts remained but were manageable; however, building refurbishing costs were estimated at well over $1,000,000 to address PCBs and asbestos-containing material.

FIG. 13.5   **E1903-11** definition of business environmental risk.

> 3.1.6 *Business environmental risk:* A risk which can have a material environmental or environmentally-driven impact on the business associated with the current or planned use of a parcel of commercial real estate, not necessarily limited to those environmental issues investigated in accordance with this practice.

business environmental risk. The nature of the transaction, the risk tolerance of the purchaser, the purchase price, and many other factors influence the selection of the appropriate scoping strategy. Today, the consultant must be aware of the needs of the marketplace and structure the scope of Phase II ESAs in a manner that affords the client sufficient information to reach business decisions.

As the industry matured through the late 1990s, many properties were identified with various types and various levels of contaminants. When it became obvious that cleaning up the contamination to "original" or to below-detection limits was impractical if not impossible, the risk-based corrective approach became more accepted at the federal and state level. With limited cleanup funding available, many properties were sitting in limbo awaiting further regulatory action. Some properties ended up with institutional or engineering controls. For these properties to be sold or redeveloped, the Phase II questions did not deal with whether the contamination was present but instead where it was and if the contaminant levels would affect the redevelopment of the subject property. This obviously dictated a different approach to the Phase II assessments than what would typically be associated with establishing applicable limited liability defenses. The concept is demonstrated in the case studies depicted in **Figs. 13.6** and **13.7**.

The **Fig. 13.7** case study emphasizes the value of collecting and reviewing all available information and defining a scope of work that is designed to answer a very specific question. In this case, groundwater impacts may indeed be present beneath the building. However, the bank already owns the property (so there is no acquisition), and there are no identified historical uses of the bank property that were suspected to be a source of the

FIG. 13.6   Case study with business risk objective.

> **Phase II case study in which liability protections are not an issue**
>
> A former gasoline service station was being considered for a development. For the past 15 years, the property was used only for car service and repair with the tanks backfilled with concrete. The tanks used to leak, and the contamination assessment report from 15 years ago identified soil and groundwater contamination. The property was in a state cleanup program, but because of risk ranking it was given a priority score that was unlikely to require any additional regulatory action at the property for the next 20 years. Experience with similar types of properties indicated that by the time the state program may reassess the property, the contamination is likely to naturally attenuate to the point that no additional remedial action will be required.
>
> The property was in a desirable part of the town, and a buyer was interested in redevelopment of the property. Although the buyer was provided with a letter from the regulatory agency absolving them of liability for the cleanup, they were concerned with how the contamination might affect the construction during redevelopment. The Phase II assessment was designed to evaluate the potential of encountering contaminated soils and groundwater in the portion of the property where the foundation excavation was to take place.

FIG. 13.7   Case study related to risks to persons on a property.

> **Phase II case study for risks to building occupants**
>
> An existing bank branch is included in a multiparcel area that is being considered for a redevelopment project. The prospective developer had conducted a Phase I ESA and identified three historical dry-cleaning operations within the proposed redevelopment area, including one of the parcels adjoining the existing bank branch. One of the parcels within the proposed redevelopment area was also an active gasoline station/convenience store that had a previous release from its underground storage tank (UST) system. During investigation activities related to the previous UST release, chlorinated solvents commonly used in dry-cleaning operations were discovered. The agency overseeing the UST release provided regulatory closure for the petroleum impacts. No additional investigation had been conducted to evaluate the source of chlorinated solvents.
>
> The developer was in the early stages of negotiations and had not yet engaged any Phase II investigations. The proposed development deal fell through, and the existing bank was now aware of an undefined groundwater plume of chlorinated solvents that extended beneath its building. The bank property was not a source of the contamination, and there was no regulatory obligation for the bank to cover the cost of collecting groundwater samples to confirm the contamination was present. Instead, a Phase II scope of work was developed that consisted of sub-slab, indoor, and background air samples to evaluate the potential for impacts to indoor air.

chlorinated solvent (so the bank would not be considered a "responsible party"). The only objective for the bank was to ensure there were no risks of exposure to the contaminants by building occupants.

E1903 does not specify which methodologies should be used. Many different standards or methodologies may be applicable. This is the area where the users are looking for help from the consultants in determining what methodology will bring out the information the users are seeking. To choose the right method, the assessor must know the extent of users' objectives and possess sufficient training and experience to implement the ESA.

The objectives set during the scoping process should enable the assessors to draw conclusions about the contamination of the subject property. The conclusions could simply be that the site has contamination, or that there is no reasonable basis to suspect that the site is contaminated. Alternatively, the conclusions could relate to the potential impacts on the expected future uses or redevelopment of the property. For the conclusion to satisfy users' requirements, the assessors must fully understand users' needs before finalizing the scope of the Phase II ESA.

As was already pointed out, in many cases establishing LLP defenses is not the objective of the user. **E1903** recognizes that in many situations the extent of the assessment will be based on the business objectives of the user. The assessor must be cognizant of these objectives when establishing the scope of Phase II ESAs with the users. **Figure 13.8** shows key sections of the standard that summarize the concepts.

Risk tolerance varies from user to user. Some users are far more risk tolerant than others. The amount of risk tolerance will also affect the scope of a Phase II ESA. When multiple users are involved in the transaction, the consultant must identify the key user and tailor the scope of the Phase II ESA to meet those needs. For example, the purchaser may be considerably more risk tolerant than the lender. Although they are both users, the ESA must satisfy the needs and risk tolerance of the lender. Failure to do so will result in "no loan" and therefore "no sale."

FIG. 13.8   Section of **E1903** describing concept of users' needs.

> 1.4 *Needs of the User*—The *user* and *Phase II assessor* must have a mutual understanding of the context in which the *Phase II ESA* is to be performed and the objectives to be met by the investigation, i.e., the specific questions to be answered or problems to be resolved by the *Phase II ESA*. The scope of Phase II activities must be defined in relation to those objectives.
>
> 1.4.1 The degree of confidence desired by the *user* influences the scope of the investigation and the evaluation of data. More extensive testing and more iterations of sampling and analysis may be needed if the objectives require detailed conclusions with high confidence. Less testing and fewer iterations of sampling and analysis may be needed if the objectives of the assessment include only general conclusions.

13.3 Limitations of E1903 Core Principles

13.3.1 ELIMINATION OF UNCERTAINTY

The principle that "no ESA can eliminate all uncertainty" applies to E1903. Invariably, the Phase II ESA is an investigation, which often includes sampling. The sample is then assumed to be representative of certain conditions of the subject property. That assumption may or may not be accurate. Therefore, assumptions impart uncertainty into the Phase II ESA. None of the ESAs, including Phase II ESAs, can eliminate all uncertainty. This is why the Phase II ESA objectives are established on the degree of uncertainty acceptable to the user. For some clients, mere detection of contamination on the subject property may be sufficient to make a business decision. Under different circumstances, the extent of the contamination may be of importance in guiding the users' business judgments. The scope of Phase II ESAs under each of the circumstances will obviously vary greatly.

The case study in **Fig. 13.9** demonstrates that the uncertainty associated with the assumptions can be reduced by increasing the level of assessment (i.e., by increasing the number of sampling locations or taking more samples in an area where a waste oil UST might be present). However, this case study also demonstrates that the law of diminishing returns applies here in that the cost of information obtained and the time required to obtain it eventually outweigh its usefulness and may hinder the timely completion of the transaction. When evaluating the adequacy of the Phase II ESA, the person performing such evaluation must consider the reasonableness of judgments and assumptions made under the circumstances present at the time of the Phase II ESA. There are many issues that can arise that can impose various limitations on the process. Selected issues are addressed below.

FIG. 13.9 Case study demonstrating uncertainty associated with assumptions.

> **Case study related to uncertainty of assumptions**
> Phase I and Phase II ESAs were conducted before purchasing an automobile dealership property. The Phase I ESA identified that six buried hoists had previously been removed from within the shop building, and one waste oil and one gasoline UST had been previously been removed. The location of the former hoists was easily identified. However, the USTs were removed before current regulations, no sampling was known to have been conducted when the USTs were removed, and there was no documentation or interview information identifying the location of the USTs.
>
> The scope of the Phase II ESA included sample locations next to each of the abandoned hoists and collection of soil and groundwater samples from each corner of the shop building (the most likely location of a waste oil tank, although not a certainty). There was no information to indicate where the gasoline UST had been located, and a geophysical survey was impractical because the property was quite large with many interferences. The Phase II sampling activities found no actionable levels of contaminants near the hoists. Relatively minor concentrations of oil and grease were identified in groundwater near one of the shop building corners. Additional investigation was conducted in that specific area, and the contaminants initially identified were found to be isolated and below regulatory action levels for commercial property. Backfill material that might indicate the location of the waste oil UST was never found; no extensive contamination was identified. The purchaser negotiated with the seller for an agreeable discount to compensate for the risks associated with the unknown gasoline UST.

13.3.2 TARGET ANALYTE DETECTION PROBLEMS (GEOLOGIC SETTING, POORLY UNDERSTOOD FATE/TRANSPORT, LIMITATIONS OF TECHNOLOGIES)

Even when Phase II ESA work is executed competently and in accordance with the standard, the consultants and users should recognize that certain conditions present especially difficult target analyte detection problems. Such conditions may include, but are not limited to, complex geological settings; unusual or generally poorly understood behavior; fate characteristics of certain substances; complex, discontinuous, random, or spotty distributions of existing target analytes; physical impediments to investigation imposed by the location of utilities and other manmade objects; and the inherent limitations of assessment technologies.

13.3.3 LIMITATIONS OR DEFECTS IN INFORMATION USED TO DEFINE OBJECTIVES

The effectiveness of a Phase II ESA may be compromised by limitations or defects in the information used to define the objectives and scope of the investigation, including inability to obtain information concerning historic site uses or prior site assessment activities despite the efforts of the user and Phase II assessor to obtain such information in accordance with Section 5.1.3.

Some users consider conducting a Phase II ESA without ever having performed Phase I assessment. If the user opts to skip the Phase I/AAI process, then they will not be eligible for any of the liability defenses, and the Phase II results will be limited. If the users are only performing the Phase II to make a business decision, then they can initiate the inquiry directly with a Phase II ESA. For example, if a user is considering the purchase of a service station with a release associated with USTs and is only concerned with the condition of the tanks, then they can hire the EP to conduct a Phase II ESA on the condition of the tanks. In effect, the user has identified a REC (a release at an underground petroleum storage tank) and is asking the EP to conduct a Phase II ESA on that specific REC.

However, this approach eliminates a very important part of the scientific method—adequate research to sufficiently define which questions need to be answered. For example, conducting a Phase II ESA for a gasoline station/convenience store without the benefit of a Phase I ESA could miss the historical fact that service activities were previously conducted or that more than one set of USTs has historically been present. It could also miss other significant sources of contamination.

13.3.4 DOES NOT DEFINE THRESHOLD LEVELS

The E1903 standard practice does not define the threshold levels at which target analytes pose a concern of significance to the user. This is where the consultant can assist the user in understanding the impact that the results will have on the user's business objectives. One common method is to apply the applicable regulatory criteria and relevant liability principles. The analysis can be expanded to meet user-defined objectives.

13.3.5 CHEMICAL ANALYSIS ERROR—BUILD IN QA/QC

Chemical testing methods have inherent uncertainties and limitations. The Phase II assessor shall build quality assurance and quality control measures into the assessment. The Phase II assessor should require the laboratory to report any potential or actual problems experienced or nonroutine events that may have occurred during the testing so that such problems can be considered in evaluating the data. The Phase II assessor should subsequently identify such problems in any reports or documentation provided to the user. Any laboratory utilized for chemical testing shall be accredited in accordance with applicable state requirements.

13.3.6 LEVEL OF ASSESSMENT VARIATIONS— COST OF ASSESSMENT VERSUS BENEFIT (EXPENSIVE DRILLING)

A common misconception among the users and assessors is that Phase II ESAs always involve sample collection. In actuality, Phase II ESAs may consist of conducting additional interviews or gathering additional historical information that can lead to a satisfactory answer to the question posed. For example, if an in-depth interview with a governmental official and a review of additional file materials reveals that the agency has reviewed the contamination on the subject property and has determined that no remedial action is required, then this information alone may be sufficient for the user to reach a decision.

13.3.7 COMPARISON WITH SUBSEQUENT INQUIRY

Sometimes information in later assessments provides a valuable hindsight. This should not be used to judge the adequacy of the findings of a Phase II ESA. The Phase II conclusions in light of the findings of a subsequent inquiry should be evaluated based on the reasonableness of the judgments that were made by the consultants at the time and under the circumstances in which they were made.

13.3.8 DATA USABILITY

Investigation data generally only represent the site conditions at the time the data were generated. Therefore, the usability of data collected as part of a Phase II ESA may have a finite lifetime depending on the application and use being made of the data. The data are sometimes used by other parties to make decisions. This should only be done after careful consideration of the initial ESA

FIG. 13.10 E1903-11 scope limitations.

> 1.5 *Limitations*—This practice is not intended to supersede applicable requirements imposed by regulatory authorities. This practice does not attempt to define a legal standard of care either for the performance of professional services with respect to matters within its scope or for the performance of any individual *Phase II Environmental Site Assessment*.

objectives. The E1903 standard cautions that, "To the extent that investigation data would fall within the scope of data used in a Phase I ESA conducted pursuant to Practice **E1527** or Practice **E2247**, the lifetime limits defined by those standards apply." The consultant should give careful consideration to whether the previously generated data are appropriate for any subsequent use beyond the original purpose for which they were collected. The data are subject to lifetime limits imposed by other laws, regulations or regulatory policies, or changing subsurface conditions.

13.3.9 ASSESSORS SHOULD NOT PROVIDE LEGAL OR BUSINESS ADVICE

The Phase II ESA is intended to develop and present sound, scientifically valid data concerning actual site conditions. The standard directs that, "It shall not be the role of the Phase II assessor to provide legal or business advice." The consultants' professional code of ethics and local, state, and federal law may also impose reporting obligations on the EPs that need to be communicated to the user early in the scoping process. **E1903-11** clearly states that it is not intended to supersede regulatory requirements, as shown in **Fig. 13.10**.

13.4 Summary

In this chapter, we expanded our understanding of the wide variety of user objectives that can be met with the Phase II ESA. We presented a series of case studies and examples that demonstrate that a thorough understanding of objectives, needs, limitations, and requirements is crucial in the determination of the appropriate scope of the work.

References

[1] ASTM Standard **E1903**, Standard Guide for Environmental Site Assessments: Phase II Environmental Site Assessment Process, *Annual Book of ASTM Standards,* ASTM International, West Conshohocken, PA, 2011.

[2] FY12 Guidelines for Brownfields Cleanup Grants, U.S. Environmental Protection Agency, Washington, DC. (http://www.epa.gov/oswer/docs/grants/epa-oswer-oblr-11-07.pdf).

Chapter 14 | Contracts

14.1 Conceptual Definition of Contracts

Contracts define the relationship between the user and the consultant. A contract is an agreement by both parties that defines mutually agreeable duties and responsibilities. The contract is a disclosure document that identifies numerous issues that may arise during the course of the Phase II Environmental Site Assessment (ESA) and describes the approach that will be used to resolve those issues. The contract also describes fees and payment obligations of the parties involved. Lastly, the contract also identifies the scope of work, sometimes by a detailed description of the tasks by reference to the standards and work plans or any combination thereof. All of the issues discussed in the previous chapters on scope and application of the standards must be considered during the contracting period of the Phase II ESA.

The broad applicability of **E1903** [1] can have a significant effect on contracting. Compared with the relatively straightforward Phase I contracts, which in essence bind the parties to follow the **E1527** [2] practice, the Phase II contracts are usually more specific to the transaction. The **E1903** is also a standard practice, but as we discussed in the previous chapters, it may cover many different scope aspects that tend to be unique to the transaction at hand. These unique aspects may not be clear to all parties at the outset and may have to be clarified in the contracts. The **E1903** does not cover contractual issues in the body of the standard. Some guidance is provided in Appendix X-2 of the standard. The appendices to the ASTM Standard are not mandatory; thus, the Appendix X-2 titled *Contracting Considerations between Environmental Investigator and User* is intended to be supplemental information for the benefit of the users of the standard. The appendix alerts the reader to the numerous issues that may need to be addressed in contracts. If the issues are not adequately addressed at the outset, then the outcome of the Phase II ESA may be adversely affected and potentially lead to litigation.

In general, if the users author the contracts, the language of the contracts will favor the users. The converse is also true. The consultant or their attorneys will generally write a contract favoring their interests and protection. Neither party should sign a one-sided contract without due consideration. Unfortunately, the constraints of time and the pressures of the market sometimes influence the judgment of one of the parties. If the party happens to be the consultant, then the eagerness to land the project can cause the consultant to commit to unrealistic goals and schedules.

In this chapter, we will discuss some of the issues that the users and the consultants should consider before entering into contractual relationships. Many of these issues are often considered and discussed by both parties, but they end up not being included in the language of the contract. **E1903** does not have any specific requirements with respect to the contents of the contract. The format and the content of the contract is left entirely up to the discretion of the consultant and the user.

14.2 Reporting Obligations and Confidentiality

During the contracting stage of the process, some of the most important issues that the consultants need to communicate to the user are the reporting obligations. A requirement to report results or in some cases even observations from a Phase II ESA to a governmental entity or third party may be imposed on the consultant by various authorities. These can include statutes, regulations, common law, and professional standards. In most cases, statutory, regulatory, and common law requirements impose reporting obligations primarily on the owner, operator, or person in charge or in control of the facility or of the property being assessed.

Often times, the consultant is hired by the purchaser. The purchaser may have no obligation to report any contamination. If some contamination is discovered at levels that raise the purchasers concern about the cost of associated cleanup, then he may want to just walk away. At this point, he does not want a report because it is just throwing more money at a bad deal and he does not care about reporting obligations of the property owner or the consultant. The consultant has no contractual relationship to the property owner, but he may have an obligation to report the discovery of the contamination to a governmental agency. This could make the owner very unhappy with the consultant. Some of these issues are best addressed by contracts.

The purchaser has an obligation to get the consent from the owner for the consultant to perform the investigation on the property. The consent language, which would be signed by the property owner, can address the reporting issues and minimize the potential for misunderstandings. The rest of the contract between the user and the consultant can clarify the obligations of and protocol for both the purchaser and consultant to report to governmental entities or third parties. The contract can also specify a termination clause that can specify that if none of the contracting parties want a written report, then it does not have to be prepared. This would constitute a user-driven deviation to the E1902 standard and should be identified as such. Because the standard is voluntary, there is nothing that prohibits this type of upfront agreement about a potential deviation.

Once the contamination is discovered, it alters the dynamics and objectives of the investigation. The case study in **Fig. 14.1** illustrates the point.

Two common objectives of a Phase II ESA are (1) Is there contamination at the subject location? and (2) How much contamination is there? The first question is answered when the contamination is discovered. The second question is more difficult to answer, and the consultant may not have adequate data to answer it. If the discovery shows low levels, such as below the regulatory cleanup levels, then the consultant may conclude that the contamination is de minimis. The potential problem with this situation relates to the confidence that the consultant has in the de minimis conclusion. In a simple Phase II in which a minimum number of samples were collected, the possibility of missing the contamination can be relatively high. The question becomes whether the results are representative of a larger area that was not sampled or whether they indicate that there may be other areas on the property that have significantly higher levels of contaminants.

As we stated previously, the regulations in many states typically place the primary reporting obligation on the owner of the property. In cases in which the owner is not the user, the reporting obligations can significantly complicate the relationship. Some consultants require the owners and the users to sign a reporting agreement, which specifies the party that will report the findings to the applicable agency if contamination is discovered during the Phase II investigation. These types of agreements can prevent some disagreements that arise from interpretation of the results.

Although the Phase II ESA can be the first step toward a characterization of the subject site, the results often fall short of being representative of the conditions of the subject property. The consultant may interpret the results to be representative of a release and believes he has a duty to report the results to the third party. The user or the owner may insist that the results are not representative of the site and are caused by other factors, such as localized contamination, sampling, or laboratory error. The difference in opinion can result in further testing. On the basis of the interpretation of the results, the parties involved will argue over who should pay, not just for the additional testing, but also for the cost of involving a third party in the transaction. Because these costs can be significant, many users and owners are concerned that the consultant may overestimate the impacts and try to impose confidentiality issues onto the consultant. All of these issues are best addressed up front before the work starts.

14.3 Written Communication

Generation and disclosure of written communication is also a contracting issue. E1903 requires a written report. This dictates that if Phase II continues to its completion, then a report will be generated.

Let us assume that the consultant walks onto the subject property and begins to conduct a preliminary field screening that uncovers an unexpected and a substantially large problem of sufficient magnitude for the user to terminate the real estate transaction. The agreement between the user and the consultant should be flexible enough to enable the consultant to stop at that point and notify the user. Both parties should have a mutual understanding of how to terminate any further effort at any given point during the Phase II assessment. The contract should also define who is privileged to the information generated up to that point.

FIG. 14.1 Case study on changed dynamics scope of Phase II assessment.

Case study on changed dynamics and objectives of Phase II assessment

The consultant was hired because some of the information in the user's possession provided substantial evidence to indicate the possibility of historical contamination of the property, and the user needed greater confidence and decided that further inquiry was warranted. If the Phase II ESA did not identify any contamination, then the user was willing to stop any further investigation and proceed with the transaction.

The Phase II ESA confirmed some contamination on the site. The levels were not very high, but they did exceed the regulatory limits in some of the locations on the property. The confirmation of contamination in the Phase II ESA shut the door on the Innocent Purchaser Defense (IPD). The purchaser now had the option to negotiate with the seller a price reduction and then purchase the property and enter into a voluntary cleanup program or try to qualify for the Bona Fide Prospective Purchaser (BFPP) under a Brownfields program.

All of these options raised new questions that needed to be answered, and they had to be answered to a higher degree of certainty than was the initial intent of the Phase II ESA. If the users chose the BFPP, then complete disclosure of all results is typically desirable. Because the BFPP carries with it continuing obligations, the purchasers will typically use the Phase II ESA to document as many existing contaminants at the site before the purchase. This can protect them in the event that there is an investigation in the future that reveals contamination not discovered before the purchase. If the purchaser did not cause the newly discovered contamination, then they may have to defend themselves against claims that they failed to meet their continuing obligations.

Because the contract clearly defined the end point of the initial assessment, it was possible for the owner and purchaser to negotiate a second iteration of the assessment that provides a more detailed site characterization. The parties agreed to split the cost of the second iteration of the assessment with the option of renegotiating the purchase price after it was completed.

14.4 Subcontracting

The Phase II assessments often involve subcontractors. The consultants should pay close attention to the subcontracting issues to ensure consistency between main contracts and subcontracts.

Portions of Phase II ESAs are often subcontracted to other parties, such as investigators, drillers, haulers, and disposal facilities. Under some circumstances, the user or the property owner may want to exercise control over the selection of the subcontractors. These issues can be problematic in that the control over selection of subcontractors will involve new potential liabilities that need to be appropriately covered in the contract language. If the user selects subcontractors to perform work for the consultant, then who is liable for the work they perform? If the consultant or the property owner selected the subcontractor, then who is liable for the work of the subcontractor? The case study in **Fig. 14.2** illustrates just some of the complexities that may be involved in subcontracting.

14.5 Generation of Waste and Exploration Damage

The case study shown in **Fig. 14.2** raises some of the basic subcontracting issues as well as the issue of generation of waste. Many cleanup operations require the consultants to consider the disposal issues. In the case of remediation, the wastes sometimes have to be removed from the site. Although many Phase II ESAs do not involve remediation, many of the exploratory activities, including a simple purge of a monitoring well, can result in generation of waste. Some of these issues can become quite complicated because they involve exploration operations on a property that is often owned by someone different from the user. The owner might object to the consultants disrupting the property and potentially generating waste. The classification of the waste materials is often not known until the results are received from the laboratory or from field tests. The consultant must consider the waste generation issues, including applicable regulatory requirements, and select appropriate techniques to minimize the impacts from the Phase II ESA exploration activities. The consultants should have upfront agreements on who owns any generated waste, who will sign waste manifests, and who will pay for the disposal.

As shown in **Fig. 14.3** even a simple direct push investigation that included core sampling and groundwater tests resulted in generation of some waste. The waste was collected in a 55-gal drum and disposed of after the laboratory results were received.

In addition to the difficulties associated with the potential generation of waste, the consultant must also consider a potential for damage because of the exploration activities. Drilling and excavation activities involve the invariable risk of damaging utility lines and underground structures. Many states have statutory obligations for contacting utilities through utility clearance programs. Coordination with utility companies or locator services should also be addressed in the contract.

Boring holes through confining soil layers can also create new pathways for pollutants to migrate to a new location. Although it may be a noble idea to demonstrate that the contaminants are confined, the consultant may actually cause the pollutants to be released into a new location because of the exploration activities. Many of these issues can result in significant impacts to third parties. The owners and operators of the facilities may have to be included in the contractual issues and work plan development to clarify the liabilities for the potential damage, accidents, and spread of contaminants.

FIG. 14.2 Case study on subcontracting.

> **Case study on subcontracting**
>
> A national environmental service firm had a master service agreement (MSA) with their client covering of the client's facilities within the USA. The MSA was structured primarily for site assessments, but it allowed for other services on an as-required basis. As part of the "other services," one of the projects included the removal, transportation, and disposal of wastewater and petroleum-contaminated sludge contained in several large aboveground storage tanks. The local office of the environmental firm subcontracted the project to a local company that operated a disposal facility near the site.
>
> The disposal facility ran into trouble, and the U.S. Environmental Protection Agency (EPA) initiated an emergency response/removal action at the disposal facility. As part of its cost recovery efforts, the EPA identified the environmental service firm as a Possible Responsible Party (PRP).
>
> The MSA with the client was based on the client's standard form of agreement, with contract terms strongly slanted in favor of the client in terms of indemnification and risk allocation. The contract between the environmental service firm and the disposal subcontractor did not include a limitation on the scope of services provision and did not require that the subcontractor possess Environmental Impairment or Pollution Liability coverage.
>
> Because of the contractual deficiencies, the environmental service firm agreed to pay approximately $60,000 to EPA in connection with damages caused by the subcontractor's mishandling of the waste material from their client.

Source: Adapted from ASTM Phase II Technical Professional Training class [3].

14.6 Responsibilities

In the context of contracting considerations, both the user and the consultant must fulfill certain responsibilities. Some of the user responsibilities are contained within the body of the standard. The user is expected to participate in the development of the scope, particularly in formulating the questions that the Phase II investigation is trying to answer. Although this appears to be a relatively commonsense task, it is often poorly defined in contracts. The users are typically not skilled in determining what kind of investigation will answer their questions and perceive that the Phase II assessment will solve all of their problems. They are often surprised to hear the limitations that are inherent to the process.

Some additional responsibilities are also outlined in the appendix of the **E1903** standard. An excerpt from appendix of the standard in shown in **Fig. 14.4**.

FIG. 14.3 Waste generation from groundwater purging in a bucket (right insert) and core sampling (left insert) collected in a waste storage drum inside of a fenced area.

Source: Photo by Zdenek Hejzlar.

The consultant also has many responsibilities, some of which are sometimes addressed in the contractual language. The foremost responsibility of the consultant is to conform to the requirement of the **E1903** Standard. If deviations from the standard are necessary, then the consultant must provide appropriate reasoning in support of the deviations. All deviations must be appropriately documented. These items are specifically addressed in the appendix of the standard. The excerpt is shown in **Fig. 14.5**.

The consultant may encounter numerous situations during the Phase II ESA that may affect the scope of the investigation. New issues may also be uncovered that were not known or previously identified. With respect to the requirements of **E1903**, the consultant should promptly notify the user of any such developments. With the current cell phone technology, keeping the client apprised of latest developments is a relatively simple task. Photographs and

FIG. 14.4 Excerpt from the appendix of **E1903** regarding user responsibilities.

> X2.8 *Responsibility of User*—The following list of responsibilities of the *user* is not intended to be exhaustive:
>
> X2.8.1 The *user* should provide access to appropriate areas of the site for the *Phase II assessor*.
>
> X2.8.2 The *user* should provide the *Phase II assessor* a site contact name and phone number.
>
> X2.8.3 When the property to be accessed is not owned by the *user*, the *user* should make provisions for the restoration of landscaping damaged by investigation activities, whether through the specifically contracted services of the *Phase II assessor* or not.

FIG. 14.5 Excerpt from the appendix of the **E1903** standard regarding the responsibility of the environmental assessor.

> X2.9 *Responsibility of Environmental Investigator*—The following list of responsibilities of the *Phase II Assessor* is not intended to be exhaustive.
>
> X2.9.1 The *Phase II assessor* should conform to the precepts of this practice and accepted industry practice. The *Phase II assessor* should document and explain significant deviations.
>
> X2.9.2 The *Phase II assessor* should provide the *user* prompt notice of environmental conditions observed.
>
> X2.9.3 The *Phase II assessor* should communicate to the *user* limitations resulting from any time and cost constraints imposed by the *user*.
>
> X2.9.4 The *Phase II assessor* should verify with the *user*, prior to implementation of the scope of work, any substantive deviations from the scope of work described in contract documents.
>
> X2.9.5 The *Phase II assessor* should ascertain and observe all site health and safety considerations and regulations applicable to the activities of the *Phase II assessor*.
>
> X2.9.6 The *Phase II assessor* should provide the *user* a written statement of qualifications, including the qualifications of the individual *Phase II assessor*(s) responsible for the *Phase II ESA* on request.
>
> X2.9.7 The *Phase II assessor* should not undertake any activity that he or she is not qualified or licensed (where applicable) to perform.

videos from the site can instantaneously be sent to keep the client or their representative fully appraised of any significant developments at the site.

The consultants need to follow any applicable site health and safety considerations, including those that may be imposed by the occupant of the property and by the appropriate regulatory agency.

Although the written statement of qualifications was required to be provided to the user in the Phase I ESA, the Phase II ESA standard requires the consultant to furnish those qualifications only if the user requests them. This is why the statement under Section X2.9.6 appears in the appendix.

Federal, state, and local regulations may require special certification or licensing requirements for the consultant to be qualified to perform certain portions of the Phase II ESA. As with any ESA, the consultant should never undertake any activity that he or she is not qualified or licensed to perform.

Although it is generally the users who impose cost and time constraint limitations on the consultant and the scope of the ESA, it is a good business practice to communicate to the users the effect of such limitations on the quality, reliability, and extent of the Phase II ESA.

14.6.1 BROWNFIELDS CONTRACTUAL CONSIDERATIONS

Many state and local governments have developed Brownfields programs to encourage reuse and redevelopment of existing properties. The programs offer various incentives for the potential developers, including special liability protection for purchasers and developers of Brownfields designated properties. These transactions involve unique consideration because they often require complex contractual agreements between the parties involved.

A significant change as a result of the EPA All Appropriate Inquiries Final Rule (AAI) changes relates to the reliance on the reports and results. Although a contract may specify that only certain parties may rely on the information, the EPA recognized that these reports will be used by others that may not have been a party named in the contract. This is often the case when the property is in the Brownfields program.

14.7 Summary

In this chapter, we discussed numerous contractual issues associated with Phase II ESAs. Through case studies, we demonstrated some of the reporting and subcontracting complications that may arise from inadequate contractual arrangements with the client. We also discussed some of the contractual responsibilities of users and consultants and how some of the issues may affect the outcome of the Phase II assessments.

References

[1] ASTM Standard **E1903**, Standard Guide for Environmental Site Assessments: Phase II Environmental Site Assessment Process, *Annual Book of ASTM Standards,* ASTM International, West Conshohocken, PA, 2011.

[2] ASTM Standard **E1527**, Standard Practice for Environmental Site Assessments: Phase I Environmental Site Assessment Process, *Annual Book of ASTM Standards,* ASTM International, West Conshohocken, PA, 2013.

[3] Technical & Professional Training Phase II Environmental Assessment Process, *Annual Book of ASTM Standards,* ASTM International, West Conshohocken, PA, 2012.

Chapter 15 | Developing the Scope of Work: Work Plans

15.1   What is a Work Plan?

A work plan for a Phase II Environmental Site Assessment (ESA) is an articulation of those elements partially described in Section 5 of ASTM **E1903** [1] on developing the scope of work.

Work plans promote clarity in defining Phase II ESA objectives and transparency in communicating and interpreting Phase II ESA results. To accomplish that goal, the **E1902** standard requires the consultant to adhere to the requirements of Section 5 of the standard. Many of these standard requirements are stringent and create significant obligations for the consultant and the user.

15.2   Statement of Objectives

The consultants are required to provide a written statement of objectives of the Phase II ESA. These objectives must include the questions to be answered by the assessment. The questions must be developed between the consultant and the user. The statements of objectives are part of the written scope of work.

The standard requires the statement of objectives to identify and describe any schedule, cost, or budget limitations applicable to the Phase II ESA. This includes any activities that are part of the assessment process.

It is important to note that the **E1903** standard requires participation by the user and the consultant in developing the statement of objectives. Any predetermined limitation on the scope of investigation, the number of iterations of sampling, or other activities that bear on the scope, schedule, or cost of a Phase II ESA must be included in the statement. The users and the consultants are required to consider whether any predetermined limitations will compromise the ability to conduct the Phase II ESA. If so, then the statement of objectives must either adapt to the objectives of the Phase II ESA so that they are achievable subject to such limitations or describe the anticipated effect of such limitations on the ability of the Phase II ESA to achieve the objectives.

The consultant becomes the educator and should carefully explain the Phase II ESA process so that the user can make informed decisions and participate in formulating the objectives. In return, the user has the obligation under the standard to provide the Phase II assessor all pertinent documentation and information regarding the property's environmental conditions that are "known to" and "reasonably and practicably available to" the user. Such information can include previous ESAs; other environmental studies; and technical reports, documents, or oral histories concerning releases or disposal affecting the property. The user should disclose to the consultant any detailed knowledge of the nature of any specialized activities and operations conducted at the property that inherently pose the potential for the presence of substances on the property. These are pertinent documents to the consultants that help them to develop an understanding of the known or potential presence of target analytes at the property.

15.3   Scope of Work Plan

Using all of the information, the consultant must then develop a description of the methods and work tasks to be implemented to achieve the user's Phase II objectives.

Here, the standard provides the consultant some leeway by recognizing that a formal written description may not be required or appropriate in all circumstances. **E1903** allows the consultant to substitute the scope of work with another document that contains the same elements, such as proposal or scope of work.

Work plans provide a road map for how the testing will be accomplished including the rationale for sampling. The consultants develop a written plan for sampling based on the hypothesized three-dimensional distribution of target analytes represented by the conceptual model.

Work plans are generally considered and at least partially developed in the scoping stage; thus, they comprise one of the first components of Phase II ESAs.

Formalized work plans are generally a standard requirement in connection with all investigations under the authority of the Comprehensive Environmental Response, Compensation, and Liability Act (CERCLA) and the Resource Conservation and Recovery Act (RCRA). All of the Brownfields programs also fall into this category. The regulatory agency usually needs to approve an investigation work plan before implementation of the investigation by the responsible party. The idea is to ensure that the scope of the investigation meets the expectations of the agency, as well as those of the responsible party, and is consistent with the principles of the National Contingency Plan.

The work plan is normally much more detailed than typical scope of service agreements that are included in contracts or proposals. A proposal usually describes the scope of services in just enough detail to help the user understand what they are paying for. Many users do not require, or even want to see, the detailed methodologies of sampling and testing procedures that the consultant proposes to use to complete the Phase II ESA. Users generally assume that the professional will follow "industry standards" and hope that the scope of work is just enough, no more or no less, than that required to satisfy their objectives. For this reason, many Phase II ESAs may not include detailed work plans, and the contracts and proposals may be sufficient to outline the Phase II ESA activities. The user's needs determine whether detailed work plans are going to be included in the Phase II ESAs.

Work plans usually include a detailed description of every increment of a given work task such as the specific sampling procedure and equipment, media to be sampled, location of samples, rationale for all of the activities, sample handling, labeling and transportation procedures, field and laboratory testing methods, quality assurance (QA)/quality control (QC) measures for every step, and even contingencies in the event that unanticipated deviations or limitations are encountered along the way. It is understandable why users are not automatically furnished with all of this detail—they are as likely to get confused and frustrated as they are to appreciate the detail.

A Phase II ESA in connection with a commercial property transaction will often not involve any regulatory agency. However, the same process of developing a work plan that evolved under CERCLA and RCRA may be of value in a commercial property assessment. If the Phase II ESA confirms the presence of hazardous substances or petroleum products under conditions that represent a release, then the regulatory agency will likely become involved. Following the protocol of a work plan will help support the agency's acceptance of the Phase II ESA results, thus reducing the risk of having to duplicate the original work effort of the Phase II ESA. For this reason, the consultant may want to discuss the advantages of work plans with users and allow the user to make the choice.

A work plan is oftne furnished as a separate document after acceptance of a proposal/contract. Most consultants consider preparation of a work plan to be more involved than a scope of work outline for a proposal.

15.4 Principles of Work Plan Development

The principles of work plan development are the same principles that the consultans must consider when proposing the initial scope of Phase II ESA to the client. Many of the issues addressed in the previous chapter on contracting are applicable to work plans. Work plan development hinges on good planning of the various components of the assessment.

15.4.1 DATA QUALITY OBJECTIVES
When the consultants are identifying the questions to be answered in the assessment, they must also consider the users' data quality objectives. Data quality objectives for a Phase II ESA are qualitative and quantitative statements that are intended to clarify the investigation objectives, define the most appropriate type of data to collect, determine the appropriate locations from which to collect the data, and specify tolerable limits on decision errors. Objectives (decisions) relating to regulatory thresholds for contaminant concentrations in soil and water, for residential or industrial land uses, or to what's acceptable in terms of contaminant migration within or outside of property boundaries are examples of data quality objectives. Meaningful data will enable the user to compare the results to those conditions that will dictate key decisions relating to the Phase II ESA and the property.

15.4.2 REVIEW OF EXISTING INFORMATION
In addition to the Phase I ESA report, there may be other readily ascertainable records, reports, or documents that will help design an investigation that will maximize the usefulness of the data obtained. As we pointed out earlier, the standard requires the user to provide this information to the consultant. The consultant uses the information to evaluate the types of chemicals associated with the release and their likely distribution on the property. The information is used to develop the conceptual site model. The conceptual site model forms the basis of Phase II investigation and uses the existing information to build the initial understanding of the property.

If not already done in the initial ESA, a detailed evaluation of available geological information should be completed and the results incorporated into the work plan. Reliable information on the soil or bedrock types, stratigraphy, depth to groundwater, and groundwater flow direction will become essential for the environmental professional (EP) in selecting sampling locations and evaluating contaminant migration routes.

Special research into the likely chemical constituents associated with the target analytes, including breakdown compounds, is usually required so that the appropriate testing parameters and methods can be proposed. This is an example in which the development of the work plan becomes a professional work effort in itself, potentially requiring a significant amount of research effort.

15.4.3 LIMITATIONS
The EP must anticipate potential physical or logistical constraints to conducting the Phase II ESA investigation. Planning soil borings on a property where the building occupies 100% of the ground surface will require special considerations above and beyond those for a vacant parcel. Planning on collecting surficial soils from a paved surface using hand utensils poses logistical difficulties. The point is to know the physical condition of the site and the logistics of access in advance so that the investigation can be implemented with as little interference as possible. A preliminary site reconnaissance before work plan development is often helpful and may be required. **Figure 15.1** demonstrates site limitations associated with being able to construct sampling wells in an area of active business activities.

Site limitations can also include logistics affecting the user's schedule and economic realities. Commercial real estate transactions are often governed by purchase agreements that specify

FIG. 15.1 Site logistical and physical limitations.

deadlines for completing due diligence activities. Deadlines are a reasonable expectation and can greatly affect what can and cannot be accomplished in a Phase II ESA investigation.

15.4.4 POTENTIAL DISTRIBUTIONS OF CONTAMINANTS

Once the likely contaminants are identified, their overall distribution on the property, including the various media (air, soil, surface water, groundwater, wastes, structures), fate, and transport characteristics, must be assessed. If the point source was identified, then what are the likely dispersion mechanisms? Could the same contaminants have been released over multiple locations on the property? Are the contaminants highly soluble and therefore likely to leach downward to groundwater? Are vapors likely to be produced, and if so, are they lighter or heavier than air? Are underground utilities likely to affect the movement of contaminants? Could dust in the buildings be contaminated? This analysis refines the conceptual site model and helps to establish media, frequency, locations, and methods for sampling and testing.

It is also important to consider the reliability and the representativeness of the data. This is especially important with soil sampling because this media is not necessarily homogeneous and the laboratory result is from a miniscule part of the property under investigation. The concept is demonstrated in **Fig. 15.2**.

15.4.5 SAMPLING PLAN

The sampling plan articulates the assessment activities that are proposed for the specific media to be sampled. Typical activities

FIG. 15.2 Soil sampling representativeness.

When do we know enough?

Assume we sample a regular 100 × 100-ft grid using a typical split-spoon sampler (3" core), two samples per boring, 1-ft-long sample per 5 ft of the target zone 10 ft thick: Sample volume = 0.05 ft³ represents 50,000 ft³, i.e.,

0.0001%

Sample actually analyzed is much smaller.

Source: Provided by Jaroslav Solc.

include soil borings or push probes; installation of groundwater monitoring wells or soil vapor probes; wipe samples, etc. For each type of sample/media, the protocol for collection, from start to sample destination, should be addressed. This part of the work plan often draws on (and refers to) the extensive body of recognized industry or regulatory standard methods and procedures (U.S. Environmental Protection Agency [EPA], state, ASTM) for the particular sampling method that is intended to be used.

The sampling plan must describe the rationale for the number and locations of samples to be collected. Is a random sampling strategy or grid system to be used? For sites with general, area-wide contamination due to diffuse manufacturing, storage, or disposal practices, a sampling grid may be appropriate. The size of the grid must be small enough to address the known or suspected variability of the contaminant distribution at the site. For sites with specific areas of contamination, samples taken from the center of the suspected contamination zone may suffice. If the general extent of contamination is desired, then additional samples taken at progressively larger distances from the center can be proposed.

The cost associated with sampling should be considered. Sampling requires mobilization of equipment and people, and cost per sample versus cost of multiple sampling iterations needs to be considered. The amount of sampling should be sufficient to meet the needs of the assessment with a reasonable degree of confidence. More sampling will not necessarily provide more useful information, as is demonstrated in the typical sampling information curve depicted in **Fig. 15.3**. The incremental amount of additional information about the site may not be worth the cost associated with the extra sampling.

Certain QA/QC procedures may be addressed in the sampling program. Decontamination procedures, chain of custody for samples, written field documentation procedures, equipment or instrument calibration, and sample trip and equipment blanks and duplicates are all part-and-parcel of the sampling program.

15.4.6 CHEMICAL TESTING PLAN

As with the sampling plan, the chemical testing plan will normally draw from the extensive body of standard test methods developed through EPA and state regulatory programs as well as ASTM or other recognized industry standards. The consultant must be knowledgeable of applicable test methods or work closely with laboratory personnel who can assist in developing a testing plan for the Phase II ESA.

15.4.7 QA AND QC PLAN

QA/QC procedures are necessary to ensure that reliable, technically sound data are obtained from the investigation.

FIG. 15.3 Information curve.

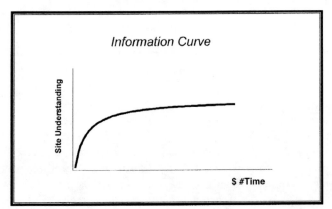

Source: Provided by Jaroslav Solc.

Documentation of the QA/QC procedures helps ensure that the data are acceptable and defensible in the event that third party or legal challenges are encountered.

QA/QC must be built into virtually every increment of the investigation. Calibration of instruments, cleaning and decontamination of sampling equipment, sample chain-of-custody documentation, and senior technical review of written reports are all examples of QA/QC procedures. Following written standard operating procedures is a means of implementing QA/QC. Testing duplicate samples, trip blanks, and laboratory blanks are common QA/QC measures for laboratory analyses.

If subcontract drilling and laboratory companies are used, then the EP should ensure that the subcontractors furnish a copy of their written QA/QC program for the specific services that they will be furnishing. Credentials, including certifications and licenses held by subcontractors, should be evaluated. A good QA/QC measure on the part of users is to evaluate the credentials of all of the EPs before entering into a contract.

Many consultants will not undertake the development of a full work plan without compensation. They recognize that development of a high-quality work plan is rendering a professional service that will likely result in an overall savings in environmental costs and reduced risk exposure to the user. A high level of expertise is required to know how to design a Phase II ESA that will achieve the user's objectives and pass regulatory muster (if necessary).

Section 5 of the **E1903** standard also includes a paragraph relating to compensation specifying that the compensation is not contingent on results of the ESA: "Payment for the Phase II Assessor's services as an independent contractor, or remuneration or job security for the Phase II Assessor as an employee of user, may not be contingent on the results or conclusions of a Phase II ESA." This was introduced into the standard to minimize the potential for conflict of interest. In some instances the user may wish or even expects certain results and may not like what the consultant has to tell them when the results come in. The standard requires the consultant to act as independent scientist and conduct the investigation and evaluate the results objectively and without reference to whether any particular outcome or conclusion is desired by the user.

15.4.8 HEALTH AND SAFETY PLAN

The consultant should prepare a site-specific health and safety (H&S) plan for the proposed investigation. The ASTM Standard does not purport to dictate what is or is not an appropriate safety plan. The H&S plan is expected to cover safety considerations for field work planned, and includes information needed in a health emergency situation, such as the location of the nearest medical facility. If a generic "boilerplate" plan is used, then it should be tailored to the investigation being conducted and the nature of the potential hazards presented at the site. Many of the H&S issues discussed in **Chapter 7** of Part 1 of this manual are specifically applicable to work plan preparation.

15.5 Summary

In this chapter, we addressed some of the issues associated with the preparations of work plans during the Phase II ESAs. We drew a distinction among proposals for a scope of work, contracts, and the detailed approach of work plans. We discussed tasks that need to be considered when preparing a work plan. On the basis of the same principles that are listed and described in the **E1903** Standard, these tasks can be presented to the user in various formats, depending on the particulars of the commercial real estate transaction and the user's needs.

References

[1] ASTM Standard **E1903**, Standard Guide for Environmental Site Assessments: Phase II Environmental Site Assessment Process, *Annual Book of ASTM Standards,* ASTM International, West Conshohocken, PA, 2011.

[2] ASTM Standard **E1527**, Standard Practice for Environmental Site Assessments: Phase I Environmental Site Assessment Process, *Annual Book of ASTM Standards,* ASTM International, West Conshohocken, PA, 2013.

Chapter 16 | Assessment Activities

16.1 Introduction

To meet the requirements of the scientific process, the Phase II Environmental Site Assessment (ESA) must include certain elements. These are generally addressed in Section 6 of the **E1903** Standard [1], and we will summarize those elements in this chapter to provide a concise overview of the Phase II ESA process. Many of the elements are discussed in further detail in other chapters, but this chapter includes the overview of the entire process and serves as a roadmap for the consultant.

The Phase II ESAs can include numerous assessment activities that are selected by the environmental professional (EP) to meet user needs. The activities can range from further literature research or interviews to field screening, multimedia sampling, and laboratory analysis. Testing is performed for detection, quantification, or both. Detection is generally an easier and less-demanding process than quantification. User needs will influence the appropriate method selection by the consultant. In the latter portion of this chapter, we will discuss some of the potentially applicable aspects of field screening and sampling activities of Phase II ESAs.

16.2 Phase II ESA Overview

In previous chapters and particularly in **Chapter 15**, we discussed the importance of establishing clear objectives. The objectives are developed by the user and the consultant. This typically involves development of a plan environmental media at a property. In most circumstances, the sampling is done in the environmental media from the locations where the target analytes are most likely to be present. Involving the user in this process is very important. Often times, multiple options are available that can be used to meet the user's objectives. The cost associated with the options can vary greatly. It is helpful for the consultant to review the different approaches with the client and discuss the benefits and the limitations of each approach. Depending on the objectives, the process may also involve third parties such as owners, buyers, and regulators. Third parties may impose constraints on the sampling process that could have a significant impact on the costs associated with the assessment. The key to the process is to develop just enough data and information regarding the target analytes to meet the objectives of the user. An additional aspect that the user should be made aware of is the fact that the objectives may not necessarily be accomplished in a single round of testing, and the Phase II ESA scope of work may warrant one or more rounds of investigation to meet the objectives.

Section 6 of **E1903** provides a general Phase II ESA overview, and the general steps that are required in a Phase II ESA are shown in **Table 16.1**.

In Section 7 of the **E1903** Standard, each of the steps from formulating questions through validation of the conceptual model is discussed in further detail. ASTM provides technical professional training in Phase I and Phase II assessments. The students have the option of attending the live course or an online training session. Several figures in this chapter have been adopted from the course. **Figure 16.1** is taken from the live training course and represents the Phase II ESA leading up to the development of the conceptual model. The numbers in parentheses refer to the section of the standard where the particular section is discussed.

16.2.1 FORMULATE THE QUESTION

To formulate questions, the user and the consultant must develop a thorough understanding of users' objectives. The Phase II standard actually requires that the assessor prepares a written statement of objectives that reflects the questions. This can be a complex process that requires the consultant to formulate the hypothesis that will then be tested in the investigation.

The Phase I assessment involved the entire property. In the Phase II ESA, only a portion of the property will typically be investigated. The consultant must determine which portion(s) of the property will be investigated. To accomplish this, the consultant utilizes all reasonably ascertainable information from previous assessments and any additional information that came to light in developing the objectives of the Phase II ESA.

16.2.2 DEVELOP A CONCEPTUAL MODEL

In **Chapter 15**, we discussed the importance of the work plan and how work plans are based on conceptual models. The conceptual model considers each area where target analytes are present or are likely present. Once a release has occurred, the manner in which it moves through the environment is affected by many factors. By the time someone conducts a Phase II ESA, the contaminants may have undergone a significant amount of migration and physical

FIG. 16.1 Elements from Phase II ESA leading up to the development of conceptual site model.

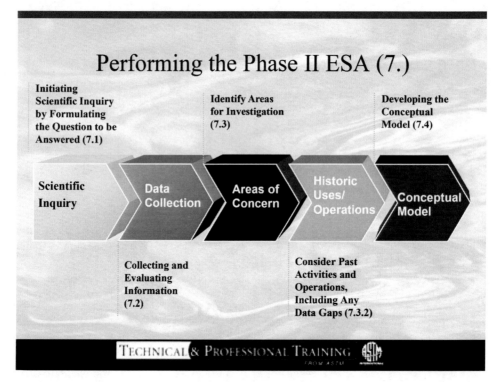

Source: Adopted from ASTM Phase II ESA technical professional training course.

TABLE 16.1 Elements of the Phase II ESA

Formulate the question
Identify the areas
Develop conceptual model
Plan sampling and testing
Sample
Validate the conceptual model
Develop conclusions and
Prepare report

changes. It is important that the conceptual model describes not only the target analytes likely to be present but also accounts for where the target analytes are likely to be located now in light of the environmental behavior, fate, and transport characteristics of the particular target analytes.

Section 6 of the **E1903** Standard identifies the necessary and required tasks that the consultant must follow to develop the conceptual model. They are listed in **Table 16.2**.

Additional helpful discussion on conceptual models is provided in Section 7 of the standard. The excerpt is provided in **Table 16.3**.

16.2.3 PLAN SAMPLING AND TESTING

In **Chapter 15**, we discussed work plans. The **E1903** Standard states that, at a minimum, the data quality objective for Phase II ESAs is

TABLE 16.2 Required Elements for Development of Conceptual Model

6.4.3.1 Identify target analytes associated with the particular substances that have, or may have, been released or may be present, based both on reported substance usage, generation, or presence and on a professional understanding of the substances typically used and generated in current and historical operations and activities.
6.4.3.2 Determine how the target analytes likely would have first entered the environment (i.e., first contacted environmental media). To do so, the Phase II assessor must draw on professional knowledge of the sorts of industrial and commercial operations and activities that are inherent to the current and historic uses of the property, known or inferred. The Phase II assessor must also draw on knowledge of the characteristics of engineered structures, features, and containers present or known or inferred to have been present at the property, from which or through which the target analytes may have been released or dispersed on the property.
6.4.3.3 Infer the environmental media and locations currently most likely to have the highest concentrations of the target analytes given the possible mechanisms of first entry into the environment; the site's physical conditions; and the behavior, fate, and transport characteristics of the target analytes, based on both known site-specific information (e.g., witness accounts of spills, location of likely releases, ground cover materials, etc.) and the Phase II assessor's professional knowledge of natural environmental phenomena and processes combined with the chemical behavior of the target analytes as well as the hydrogeology and geochemistry of settings similar to those of the property.

to achieve reproducible chemical testing results. To be consistent with scientific inquiry, the work should be formulated such that another Phase II assessor would be able to reproduce the assessment and obtain consistent results. Given the changing and dynamic environment of soil and groundwater conditions, this

TABLE 16.3 Developing the Conceptual Model from the E1903 Standard

> 7.4 *Developing the conceptual model*—For the purposes of a *Phase II ESA*, the *conceptual model* consists of a description of the likely environmental conditions of the *property* relative to the *presence* or likely *presence* of *target analytes* in *environmental media*. The model hypothesizes (i.e., predicts) where specific *target analytes* would occur now, in light of the likely mechanisms by which *target analytes* were released or may otherwise be *present*, how and where they likely first contacted *environmental media*, the environmental *behavior, fate*, and *transport characteristics* of the particular *target analytes* and/or the compounds or mixtures of which they are a part, and physical characteristics of the *site* that would influence the persistence and distribution of the *target analytes* (e.g., *transport* or *migration pathways*) should a *release* have occurred. The *conceptual model* must be conceived prior to sampling and *chemical testing* to guide the work, and must be refined throughout the investigation process to incorporate new information as the body of knowledge about site conditions evolves. The components of the *conceptual model* are described in 7.4.1 through 7.4.3, below, and include determining the *target analytes* (7.4.1), hypothesizing the mechanisms by which *substances* first entered into the environment and the points of entry (7.4.2), and hypothesizing the *behavior, fate*, and *transport characteristics* of the *target analytes* (7.4.3).

requirement of the standard could present challenges. The results of a consequent Phase II assessment may not be consistent if something changes between the assessments.

16.2.4 SAMPLE

The consultant should perform the sampling and chemical testing in accordance with the plan, making observations and note of actual physical conditions revealed by the investigation. Documented subsurface soil and groundwater characteristics encountered during testing, including any physical or logistical impediments to such physical barriers and barring sampling at specified locations or insufficient sample volume recovered, are all valuable parameters when the consultant has to interpret the data. **Chapter 7** expands further on some of the impact of this information on evaluation of the data.

16.2.5 VALIDATE THE CONCEPTUAL MODEL

By evaluating the chemical testing results and other investigation findings at the completion of the investigation, the consultant can then determine whether the available information is consistent with the conceptual model. Only a validated conceptual model can be used to support sound conclusions regarding the presence and significance of target analytes. Further detailed discussion of this topic is provided in **Chapter 17**.

16.2.6 DEVELOP CONCLUSIONS

With the validated conceptual model, the consultant is now in a position to develop the conclusions of the Phase II ESA. They are based on an interpretation of all results and findings and are consistent with the validated conceptual model. The conclusions must specifically answer the questions that the Phase II ESA set out to address. In cases in which some of the questions have not been answered, the consultant must clearly state in the report why those questions cannot be answered and what conclusions, if any, can be drawn.

16.2.7 PREPARE A WRITTEN REPORT

The preparation of the report is the subject of **Chapter 18**. A written Phase II ESA report is a required element of the **E1903** Standard. The report must include objectives, findings, interpretations, and conclusions along with descriptions of the conceptual model, the investigation(s) performed, observations made, and data obtained.

16.2.8 FIELD SCREENING AND ANALYTICAL TECHNIQUES

Field screening methods are often used during the Phase II ESA to characterize certain attributes of the site. Many of the methods and instruments are designed primarily for detection and provide limited results with respect to quantification. New, more accurate field screening instrumentation is being introduced into the marketplace, and the consultant must stay abreast of the developing technology to stay competitive and to be able to apply the appropriate technology to meet the user's needs.

The advantages of field screening methods are that they generally enable the consultant to quickly obtain data in that the results are read directly from the instrumentation at the site and do not require transport of the sample to the laboratory. This also allows multiple sampling to confirm the results, and it enables the consultant to sample multiple locations. Some of the instruments are also capable of providing some quantification data. The consultant can detect higher concentrations and use the real-time ability of field screening to pinpoint the areas of highest concentrations. Such information can be extremely useful in identifying potential sources of contamination or preliminary mapping of the horizontal and vertical distribution of contaminants. Even if the data are not sufficient to answer all of the questions, they can guide the collection of samples for more rigorous laboratory testing or field analysis.

Field screening methods can be extremely cost effective. Once the instrument is set up in the field, the cost to run several additional samples is minimal compared with the cost of laboratory analysis.

Some of the disadvantages of the field screening methods are associated with the reliability and accuracy of the data as well as the reliability of the instrumentation itself. Laboratory tests generally provide more accuracy and precision, but they need more time to process and require strict quality control during sampling, storage, and transportation to the laboratory. The consultant must consider the limitations of the field screening methods with respect to the usability of the data. If the data obtained are adequate to provide the user with satisfactory information, then the field screening methods may provide a fast and efficient way to meet users' needs. On the other hand, the EP may consider the use of mobile laboratories for selected projects.

Field screening methods are used to assess soil, soil gas, surface water, or groundwater. Field screening has also been used to detect subsurface physical anomalies, potential migration pathways, and groundwater volatile organic plumes. The methods typically use the principles of ion-selective electrodes and the detection of volatile organic compounds (VOCs) using field gas chromatographs (GCs). The detection of inorganic constituents and semivolatile organic

FIG. 16.2 Field screening using the PID at an excavation.

Source: Adopted from ASTM Phase II ESA technical professional training course.

FIG. 16.3 Ground-penetrating radar.

Source: Adopted from ASTM Phase II ESA technical professional training course.

FIG. 16.4 Example of output from ground-penetrating radar.

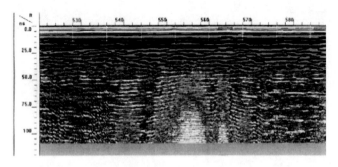

FIG. 16.5 Magnetometer survey.

Source: Adopted from ASTM Phase II ESA technical professional training course.

chemicals is accomplished using colorimetric wet chemistry methods, including enzyme immunoassay tests. Instruments include portable flame ionization detectors (FIDs), portable photoionization detectors (PIDs), GCs, X-ray fluorescence (XRF), and passive soil gas collectors. **Fig. 16.2** shows a PID used to screen soil at an excavation site for VOCs.

Field analytical techniques are not limited to instrumentation that samples media and analyzes them for potential contaminants. Geophysical methods using ground-penetrating radar or electromagnetics can also be used to identify underground structures, including storage tanks and piping. **Figure 16.3** shows the use of ground-penetrating radar. **Figure 16.4** is an example of output from ground-penetrating radar.

Figure 16.5 shows a magnetometer survey for buried metal drums. Magnetometers detect metal objects underground and can be used to locate tanks, piping, and other metal containers that may have been buried on site.

Figure 16.6 demonstrates the use of portable XRF screening soil for metal contaminants. These instruments can detect heavy metals in soils. The sensor penetrates several millimetres into soil, and the portable gun can be used directly as shown in the figure or samples can be taken from various depths placed in sample holders and analyzed in that manner.

Many of these methodologies can be useful in site screening, especially when the scope of the Phase II investigation includes screening for specific contaminants of objects or when the data are used for site characterization. Some of the screening methods can be quite complex, requiring specialized expertise to interpret the results.

Another screening method uses laser-induced fluorescence (LIF) [2,3]. **Figure 16.7** illustrates the basic principles of subsurface fluorescence screening systems. In situ tools such as LIF are capable of delineating contamination faster and with higher definition than many other field methods. LIF screening systems typically exhibit semiquantitative monotonic behavior for standard light nonaqueous phase liquids (LNAPLs) such as diesel, jet fuel, and gasoline with limits of detection (LOD) of 100 mg/kg. Some LIF

FIG. 16.6 Use of XRF portable meter on soil.

Source: Adopted from ASTM Phase II ESA technical professional training course.

systems are designed specifically for coal tars and creosotes with typically higher LOD ranging from 100 to 500 mg/kg.

The LIF sensor is deployed with either percussion or static direct-push technologies. LIF uses various lasers (dependent on the company providing the LIF or the product being delineated or both) coupled with an optical detector to measure fluorescence via optical fibers. The measurements and detection are made through a sapphire window on a probe that is pushed into the ground by a truck-mounted direct-push machine such as Geoprobe or Cone Penetrometer Test (CPT) systems.

The LIF probes are typically advanced at an ASTM Standard speed of 2 cm/s. The LIF method provides data on the in situ distribution of petroleum hydrocarbons based on the fluorescence response induced in the poly aromatic hydrocarbons (PAH) compounds. The method provides a real-time detect/nondetect field screening capability relative to a detection limit for specific fuel products on a site-specific soil matrix. The technique does not provide species-specific quantitation; however, it can produce semi-quantitative results at concentrations within two orders of magnitude of its detection limit for fluorescent fuel hydrocarbons [2].

Figure 16.8 illustrates the multiwavelength delay fiber detection system concept. The resulting waveform data provide valuable insight into the quantity and identity of the fluorescing materials. This allows characterization of the subsurface fluorescence "on the fly" during the push because the waveforms are recorded quickly (waveforms averaged each second) [3].

Another example of a field screening method is depicted in **Fig. 16.9**. AQR Color-Tec is a field-based analytical method that combines sample purging with colorimetric gas detector tubes to detect total chlorinated volatile organic halocarbon compounds in any ex situ liquid or solid sample at concentrations from approximately 3 μg/L or μg/kg. Samples are analyzed in 2 min or less by purging the volatile compounds from the sample directly through the colorimetric tube, which is designed to produce a distinct color change when exposed to chlorinated compounds. Estimated sample concentrations are obtained by comparing the tube readings to a conversion table, which was developed based on comparison of the method values to GC/mass spectroscopy (MS) analysis of split samples. **Figure 16.10** demonstrates the principle of operation.

Depending on the nature and complexity of the Phase II ESA, the consultant may also select other field testing methodology to generate useful information that helps to identify contaminant pathways or conditions of structures. For example, dyes can be used to trace drains and groundwater patterns in fractured rock, thickness gages to evaluate conditions of tanks, flow and turbidity meters to analyze discharges or stream characteristics, and video probes can be used to inspect inaccessible areas such as wells and sewers.

Field screening and field analytical methodologies are used to generate information that provides qualitative confirmation of contaminants. It answers the question with a simple yes or no. Some methods also provide information that gives some quantitative answers (i.e., How much?). For example, **Fig. 16.11** shows a soil vapor sampling for field GC testing. With the consultant's help, the users can evaluate the information together with risk tolerance and determine whether the results are sufficient to make business decisions.

In the past decade, the advances in instrumentation have made it possible for selected laboratory methods to be completed in a mobile laboratory at the site. When quick turnaround times are essential, it may be appropriate to bring a state-certified mobile laboratory onto the site to analyze samples as they are collected. This can provide a high level of accuracy with quick turnaround. **Figure 16.12** is an example of such a laboratory.

16.3 Environmental Media Sampling

The **E1903** Standard does not discus or specify any specific environmental sampling methods or field or laboratory analysis of strategies. These decisions are left up to the consultant. The standard does require the consultant to provide the rationale for all of the strategies and test methods used in the process. For this reason, documenting the thought process becomes important. The assessment activities are designed to determine whether hazardous substances and petroleum products or other conditions of concern to the user are present on the property. The consultants specify which activities will take place in the work plan and select sampling location and analytical parameters that focus on achieving the objectives agreed upon with the user. Many ASTM Standards provide further guidance with sampling methodology and protocols. Some include detailed descriptions with figures and tables to assist the consultant in performing the required elements. For example, **Fig. 16.8**, taken from **D6001-12** *Standard Guide for Direct-Push Groundwater Sampling for Environmental Site Characterization* [4],

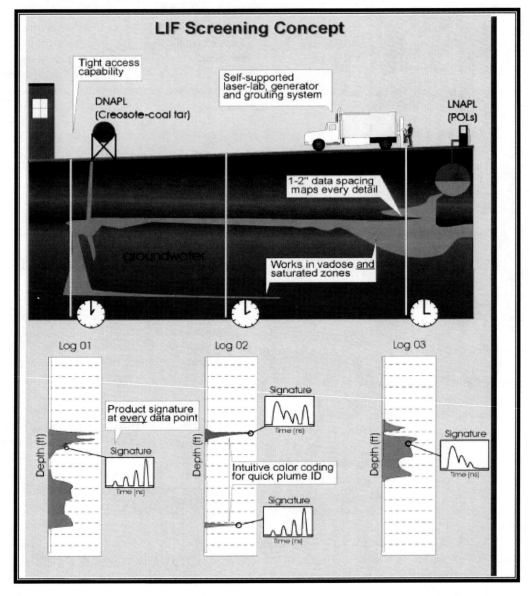

FIG. 16.7 Laser induced fluorescence.

Source: Dakota Technologies, Inc., 2201-A 12th St. N., Fargo, ND 58102.

shows a simple protected screen sampler. Under some circumstances, the sampling procedure will require some specific procedures to stabilize the sampling media. For example, **D4750** *Standard Test Method for Determining Substrate Liquid Level in a Borehole or Monitoring Well* [5] describes procedures for measuring liquid levels in establishing a stabilized condition at a sampling location.

E1903 requires the consultant to provide sufficient documentation of the sampling process so that if the process was performed by another consultant using the same methods and test methods, then the results would be consistent. The documentation of the process can be a key element for being able to accomplish this goal.

The consultants must sufficiently document the sampling locations. Depending on the objectives of the assessments, the sampling location documentation can range from general identification of a sampling location on a sketch to detailed surveyed locations. GPS location devices are becoming increasingly popular and relatively accurate tools in developing location documentation.

16.4 Sample Handling

Sample handling includes all of the activities during sample collection, containerization, filtration, preservation, and transportation to the laboratory. Often even simple quality assurance measures such as having a nice meniscus during groundwater collection to avoid volatilization of the sample in air bubbles can affect the results. **Figure 16.13** shows the desired meniscus before placing

FIG. 16.8 Multiwavelength delay fiber detection system concept.

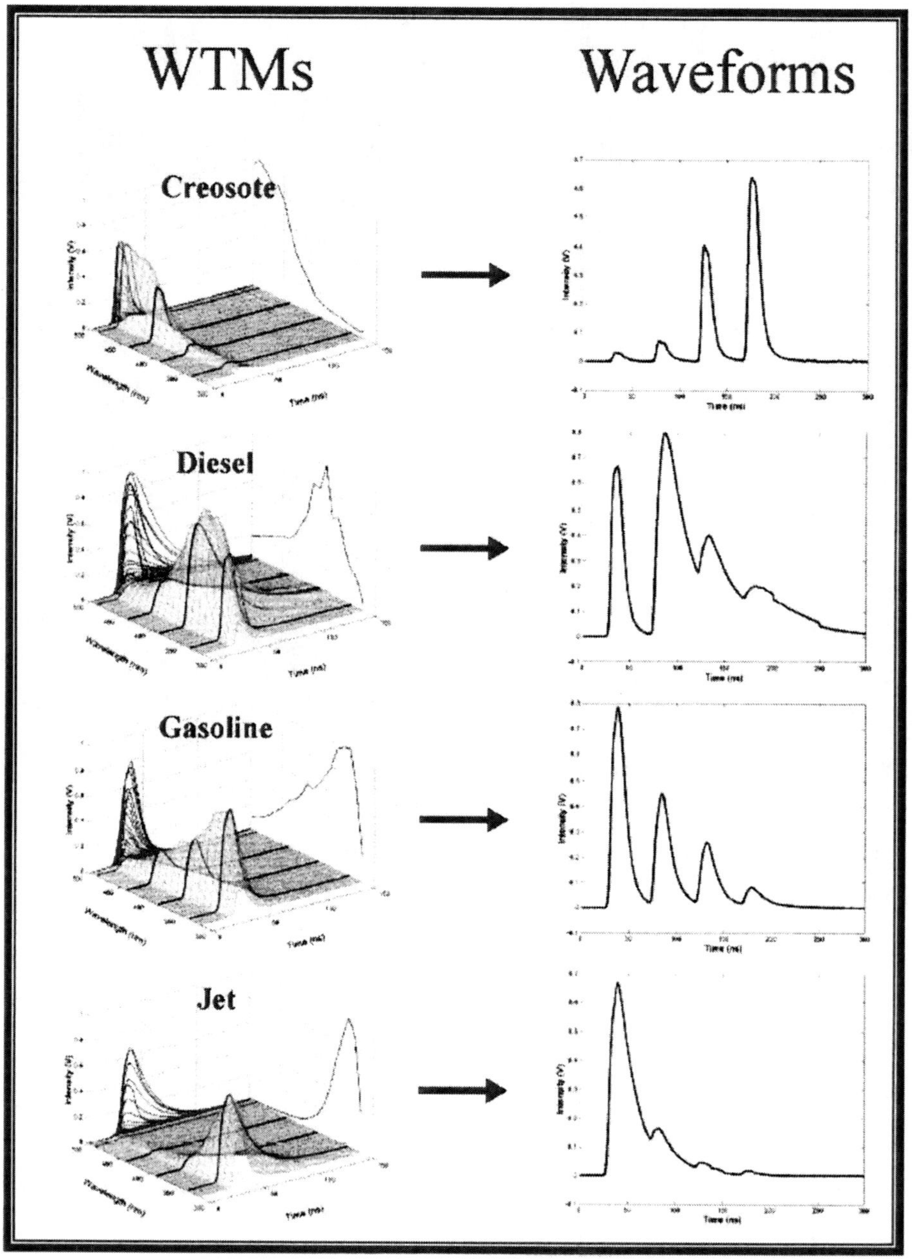

Source: Dakota Technologies, Inc., 2201-A 12th St. N., Fargo, ND 58102.

a cap on the sample. **Figure 16.14** shows flow-through cell and low-flow sample collection.

In addition, once the samples have reached the laboratory, they may be handled further in preparation for the specific testing. Certain tests require only a portion of the submitted sample to undergo testing. The laboratory is actually taking a sample of the sample. Certain homogenization procedures may be used in an attempt to ensure that the smaller sample taken is an accurate representation of the larger sample. Under some circumstances, the sample handling should be minimized. For example, soil samples collected for VOC analysis should be handled as little as possible and homogenization is prohibited by standard procedures.

The consultants must consider the potential for contamination during sampling activities. Contaminants can be introduced into the sample from the sampling equipment and other environmental factors. Every step in the sampling process can potentially cause contamination of the sample. Multiple sampling is susceptible to cross-contamination and documentation errors. Adequate care and appropriate quality control measures should be taken to

FIG. 16.9   Color-Tec field screening for chlorinated solvents.

Source: Adopted from ASTM Phase II ESA technical professional training course.

FIG. 16.11   Field GC testing of soil gas samples.

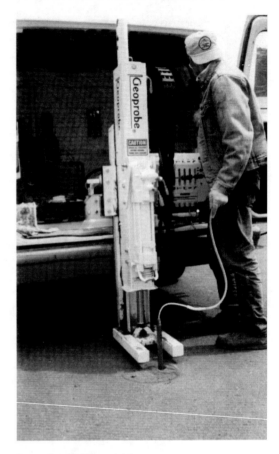

Source: Peer Environmental, Inc.

ensure the integrity of the process. **Figure 16.15** shows documentation of the hole and documentation of sample collection.

Drilling and excavation procedures in contaminated media have the potential to spread the contamination or generate new pathways for migration. The consultants must take adequate care to minimize these factors. New sampling technologies and equipment specially designed to minimize potential errors are being introduced into the marketplace on a frequent basis and require the EPs to stay current on technological improvements and innovations.

Depending on the media being sampled, the consultant will have to utilize applicable standardized methodology for preserving and transporting the samples. **Figure 16.16** depicts a shipping box design for soil sample transport. The figure is reprinted here from ASTM **D4220** *Standard Practices for Preserving and Transporting Soil Samples* [6]. This standard was primarily intended for soil samples that are to be tested for engineering properties, but the principles outlined are applicable for samples of soil for other evaluation purposes, including potential contaminant-containing soils. Organic chemical analysis testing often requires the samples to be maintained at uniform temperatures or refrigerated. For this reason, many sample handling procedures utilize coolers to transport samples to the laboratory.

The transportation of samples may include shipment of samples by commercial carriers. Transportation, containment, storage, and disposal of samples obtained from contaminated sites may be subject to regulations established by federal, state, and local agencies.

FIG. 16.10   Method overview for AQR Color-Tec.

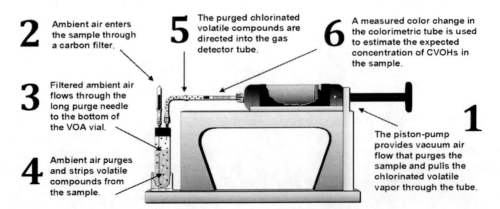

Source: www.aqrcolortec.com.

FIG. 16.12 Interior of truck-mounted mobile laboratory.

Source: Analytical Laboratories of Florida; photographed by Zdenek Hejzlar.

FIG. 16.13 Meniscus on a volatiles sample bottle.

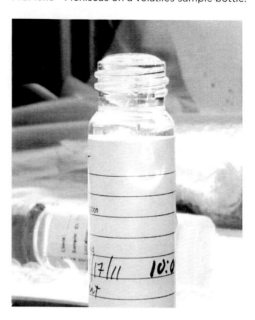

The chain of custody record is a formalized written documentation providing information about sample collection and handling. It identifies dates and times of collection and transfer among all individuals involved in the chain of sample possession. It also identifies the requested testing, generally by listing or referencing an agency procedure. **Figure 16.17** is an example of a chain of custody record. Most laboratories provide chain of custody forms that are designed for the type of testing and analysis that they perform and make those available to consultants.

The completed forms are a required part of the written report documentation of **E1903** and are generally included as appendices. The completed forms also provide information about the time and date at which certain activities were completed. Some methods require specific sample holding times that should not be exceeded. **E1903** recommends that the EP take appropriate measures to deliver the samples, enabling the

FIG. 16.14 Flow-through and low-flow sample collection.

FIG. 16.15 Documenting borehole and sample collection.

FIG. 16.16 Shipping box design for short tube or ring samples.

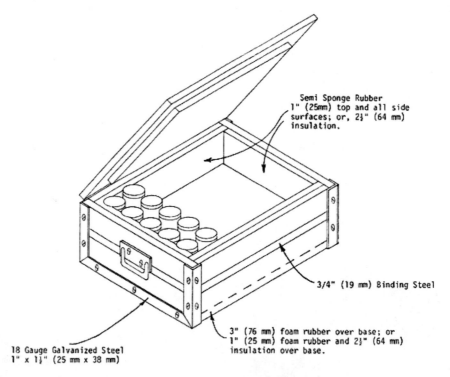

Source: From ASTM **D4220.**

analysis to be completed within the appropriate sample holding times.

16.5 Summary

In this chapter, we discussed issues associated with assessment activities. We considered some of the advantages and disadvantages of field screening compared with laboratory analysis. We learned that Phase II ESAs may include field screening and laboratory analysis. We identified the vast array of ASTM and non-ASTM Standards and referenced other documents that may become applicable or useful materials for the assessment activities portion of Phase II ESAs.

FIG. 16.17 Example of chain of custody form.

Sample Identification/Traceability Record
(Controlled Document)

Project: ______________________ W.O. * ______________
Shipped by: ______________________
Shipped to: ______________________ Attention of:
Comments: ______________________ Hazardous materials suspected? (yes/no)

Sampling Point	Location	Field ID *	Date	Sample Type	No. of Containers	Analysis/Test Required	(optional) Lab ID

Sampler(s) (signature) ______________________

Field ID	Relinquished by: (signature)	Date/Time	Received by: (signature)	Date/Time	Comments

Shipment prepared by: (signature) ____________ Date/Time ________ Shipment method: ________
Received for Lab by: (signature) ____________ Date/Time ________ Comments ________

Receiving Laboratory: Please return original form after signing for receipt of samples.

Source: From ASTM **D4220**.

References

[1] ASTM Standard **E1903**, Standard Practice for Environmental Site Assessments: Phase II Environmental Site Assessment Process, *Annual Book of ASTM Standards,* ASTM International, West Conshohocken, PA, 2011.

[2] Stepan, D. J., Sorensen, J. A., and Solc., J., "Real-Time In Situ Detection of Organic Contaminants by Laser-Induced Fluorescence System," Energy & Environmental Research Center, Grand Forks, ND, 1999.

[3] St. Germain, R., Gillispie, G., Adamek, S., and Rudolph, T., "In-Situ Optical Screening Tool for Subsurface Coal Tar and Creosote NAPL," Dakota Technologies, Inc., Fargo, ND, 1999.

[4] ASTM Standard **D6001**, Standard Guide for Direct-Push Groundwater Sampling for Environmental Site Characterization, *Annual Book of ASTM Standards,* ASTM International, West Conshohocken, PA, 2012.

[5] ASTM Standard **D4750**, Standard Test Method for Determining Substrate Liquid Level in a Borehole or Monitoring Well, *Annual Book of ASTM Standards,* ASTM International, West Conshohocken, PA, 2001.

[6] ASTM Standard **D4220**, Standard Practice for Preserving and Transporting Soil Samples, *Annual Book of ASTM Standards,* ASTM International, West Conshohocken, PA, 2007.

Chapter 17 | Validation Process and Interpretation of the Results

17.1 Introduction

The assessment activities undertaken during the Phase II Environmental Site Assessment (ESA) generate data that need to be critically analyzed by the consultant before the consultant can begin to interpret the results with respect to the contamination of the subject property. The first step in this process is to evaluate the quality assurance (QA) and quality control (QC) information. This provides the consultants with insights into whether they can trust the numbers before proceeding to interpreting the results. Once we can trust the numbers, we can use them to validate the conceptual model and proceed to the last step of interpreting the results. In the standard, data validation and data interpretation topics are covered partially in Section 7 and Section 8. In Section 7.5.7 of **E1903** [1], the standard discusses some of the QA and QC aspects of the ESA.

17.2 Evaluation of Data

The standard does not require the QA/QC plan to be in the form of a written document, but it does require that the consultant have knowledge of them and follow them. Having good documentation helps when the time comes to evaluate the data.

The QA of the assessment is the management function related to sampling. It considers the decisions related to the sampling plans that are based on the conceptual site model and the underlying assumptions about the potential distribution of contaminants on the property. The QA plan ensures that the samples collected are representative of the environmental media and locations consistent with the conceptual site model. Many of the aspects of QA such as sampling locations and parameters rely on the conceptual model assumptions. Therefore, during the validation process when the consultants are verifying the QA aspects of the data, they are also evaluating the applicable portion of the conceptual model.

The QC of the assessment is more of the line function that specifies the testing criteria that need to be followed by the field personnel to meet the goals of the QA plan. It ensures that sample integrity is not compromised with regard to target analyte presence and levels as a result of the sampling and sample handling procedures. The QC function also specifies the appropriate sample handling procedures before delivery to the laboratory, and it ensures that samples are properly preserved (e.g., refrigerated, or combined with appropriate preservative chemicals). It sets schedules so that the chemical testing can be accomplished within required holding times and that sample chain of custody is properly documented. The QC function also sets procedures to ensure that deviations from the sampling plan are adequately documented. This approach ensures that the data and information developed are accurate, reproducible, and consistent with normal requirements for scientific inquiry.

Certain QC steps during the sampling process, such as documentation of sampling parameters, blanks, and duplicates, can help in validation of the data, but they increase the cost associated with the collection and analytical portion of the Phase II ESAs. The cost of the QC steps is sometimes challenged by the clients because they do not perceive any added value by doing and having to pay for these steps. Nonetheless, the additional cost can be considerably lower than having to mobilize and retest again. Many consultants discuss QA/QC procedures with the user to make them aware of the effect of various QA/QC decisions.

Once the QA and QC information has been verified, the consultant can proceed with validation of the conceptual model. In this step, the consultant systematically evaluates the data and problems or questions being addressed in the assessment. This step is done to determine whether a sound and sufficient understanding of site conditions has been gained.

In the validation step, the consultant must question if the conceptual site model predictions of target analytes' points of entry, migration pathways, and current distribution are supported by the results of the investigation. If the data support the predictions, then the consultant can conclude that the conceptual model was validated.

The **E1903** standard points out that "the information and data often do not speak for themselves, and must be analyzed to qualify the sufficiency of the investigation." This is an important point that was added to reiterate the importance of the Phase II report in which the validation process is documented. Often times the client just wants the result and does not want to pay the additional cost associated with preparing a report in which the consultant would go through the validation. The importance of the validation step was one of the reasons the committee made the report a mandatory part of the process. The conceptual site model validation

FIG. 17.1 Validation process from ASTM **E1903-11** [1].

> 7.7.1 *Validation of Assumptions*—The first task in analysis of data is to consider whether the assumptions upon which the *conceptual model* and sampling plan were based were valid. That is, one asks whether samples were collected of the appropriate *environmental media* (for example, soil or groundwater) at the right location and depth (for example, where the highest concentrations of *target analytes* should be, if a *release* had actually occurred).
>
> 7.7.1.1 The *conceptual model* is usually based initially upon assumptions regarding subsurface physical conditions, such as relative permeability, depth to the *water table*, and *groundwater flow direction*, as well as regarding the nature of the *target analytes* (e.g., the form and volume of a hypothesized *release*). Actual information on these conditions is usually gained through the explorations, sampling, and observations of the investigation (at least for those studies where subsurface conditions are of concern). Upon completion of a round of sampling and testing, the *Phase II assessor* must evaluate whether the assumptions were valid in light of the actual conditions encountered. If an assumption was not valid, then the investigation may not have accomplished its purpose and additional iterations may be warranted.
>
> 7.7.1.2 If the results of the latest round of investigation are insufficient to support a validated *conceptual model*, determine whether an additional iteration of investigation is warranted. Any iteration should begin with a re-evaluation, by the *Phase II assessor* in consultation with the *user*, of the question(s) being addressed (6.4.1) and the cost and likely effectiveness of additional iterations. Depending upon that re-evaluation, additional iterations of investigation may involve repeating some or all of the component steps of a *Phase II ESA* (6.4.1 through 6.4.5) in order to support a validated *conceptual model* (6.4.6) before formulating the *Phase II ESA* conclusions (6.4.8).
>
> 7.7.2 *Validation of Chemical Testing Data*—The second task in analysis of data is to evaluate whether the quality of the *chemical testing* data is satisfactory according to the *QA/QC* plan (7.5.7). For example, any *target analytes* detected in the samples must be confirmed to be, in fact, attributable to the *presence* of *substances* at the site, rather than to sampling, handling, and testing artifacts. Similarly, the reported absence of detectable *target analytes* from samples must be evaluated against the detection limits and any surrogate recoveries achieved in the analyses. When data quality is determined to be acceptable, investigation results may be interpreted.

requires that the consultant determine that the investigation results are consistent with the site model that has been updated with all of the information gained through the investigation. The consultant then can conclude that the investigation has been adequate to understand site conditions and resolve the questions that were being addressed. **Figure 17.1** is an excerpt from the **E1903** standard that itemized the validation process.

Sometimes the consultant may get involved in a project in which they did not conduct the Phase II investigation but were called upon to help understand and interpret the data. The following case study illustrates this point.

During a Phase I assessment, the consultant that was hired by the prospective purchasers raised an issue with a neighboring property. The subject property was undeveloped land in an industrial zoned area adjacent to a refuse transfer station. In their analysis, the consultant theorized that the leachate pit one-quarter of a mile from the subject property was likely to be contaminating the subject property and recommended a subsurface groundwater Phase II investigation along the boundary of the subject property. The investigation consisted of a direct-push investigation in three locations along the boundary of the property. The ground was saturated at the time and water was encountered at 0–6 in. below the surface, and the samples were taken from a screened portion from 6 in. to 3 ft below the surface. The samples were sent to a laboratory for analysis. The results from one of the three sampling points came back with elevated levels of ammonia. The consultant concluded that the subject property was impacted by the leachate, as suspected.

The consultant did not include or perform any groundwater QC information parameter testing such as pH, turbidity, dissolved solids, etc., stating that it was not necessary for this kind of investigation. Purging information was also not documented. **Figure 17.2** is an example of typical data collected in direct-push sample collection that was omitted by

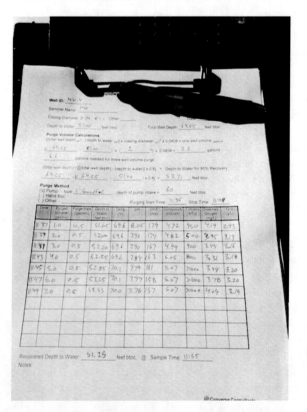

FIG. 17.2 Sampling QC parameter documentation.

FIG. 17.3 Sampling parameter instruments.

Source: Photograph provided by Norman Eke.

the environmental professional (EP) in this investigation. **Figure 17.3** shows the typical instrumentation used to generate the parameter information.

At this point, the owner of the property hired their own consultant to get a second opinion. Because of the omission of this QC step during sampling, an independent third-party verification of the data could not be done. The subsequent review of the laboratory results revealed that although the ammonia was present, the rest of the results did not resemble a typical leachate fingerprint, but rather that of cattle urine. Subsequent investigation by the owner's consultant revealed that cattle were occasionally brought onto the property, resulting in relatively localized and temporary elevation of ammonia in surface water. The sampling technique of the EP caused the surface impacts to be drawn into the sampling point. The timing of the investigation caused the prospective purchaser to lose his $50,000 deposit and to lose the sale to another purchaser. The prospective purchasers sued their consultant for loss of the deposit and the loss of the sale.

17.3 Interpretation of Results

After the validation process has been completed, the consultant can then interpret the Phase II ESA results to determine the significance of the data as they relate to the objectives of the user. The steps of the results' interpretation are typically documented in the report. The interpretation of the results is addressed in Section 8 of the **E1903** standard.

When the presence of target analytes is confirmed at the site during the investigation, then the interpretation details may depend on the user's objectives. The consultant may interpret the results in relation to numerical criteria, including regulatory criteria if relevant to such objectives, or in relation to other quantitative or qualitative criteria based upon the user's needs or contractually imposed conditions. When evaluating chemical testing results against any such criteria, it is essential that the consultants compare results to the conceptual model. It may be difficult to determine if the results are actually representative of site conditions. The consultant should consider whether the results represent conditions of a particular area or if they represent spots of the highest concentration. The standard points out that it would be inappropriate for the consultant to dismiss the target analytes associated with a release on the basis that target analyte levels do not exceed relevant regulatory levels unless data represent the highest levels. The interpretation of results will likely fall into one of the following categories.

17.3.1 NO-DETECT RESULTS

If the Phase II investigation conformed to the standard and target analytes were not detected at concentrations above laboratory reporting limits appropriate to the objectives of the assessment, then the consultant can render an opinion that there is no longer any reasonable basis for believing that target analytes are present.

Often times the scope of the investigation is limited by the user to specific target analytes. Other contaminants may be present that the user did not want investigated. The standards committee recognized these types of limitations and permits the consultant to conduct the investigation under these constraints. However, the standard cautions the consultant to clearly qualify and state in the report that the conclusion "no further inquiry warranted" applies only to the releases and likely release areas assessed.

17.3.2 LOW-LEVEL CONTAMINANTS RESULTS

There are instances in which the contamination was detected at some low level that is within the user's risk tolerance level. This is particularly applicable when establishing Landowner Liability Protections (LLPs) under the Comprehensive Environmental Response, Compensation, and Liability Act (CERCLA) is not the user's objective. In these cases, the consultant may eliminate a likely release area from further assessment based on a Phase II ESA finding that there is no reasonable basis for believing that target analytes are present at levels that are of concern in light of the objectives of the assessment.

17.3.3 DEALING WITH UNEXPECTED RESULTS

In some cases, the investigation reveals unexpected results and the Phase II ESA fails to achieve the objectives of the assessment. The standard still requires the consultant to interpret the results in relation to the objectives. Any difference between objectives and results must be described and evaluated by the consultant. The standard requires the Phase II assessor to determine the need for and scope of

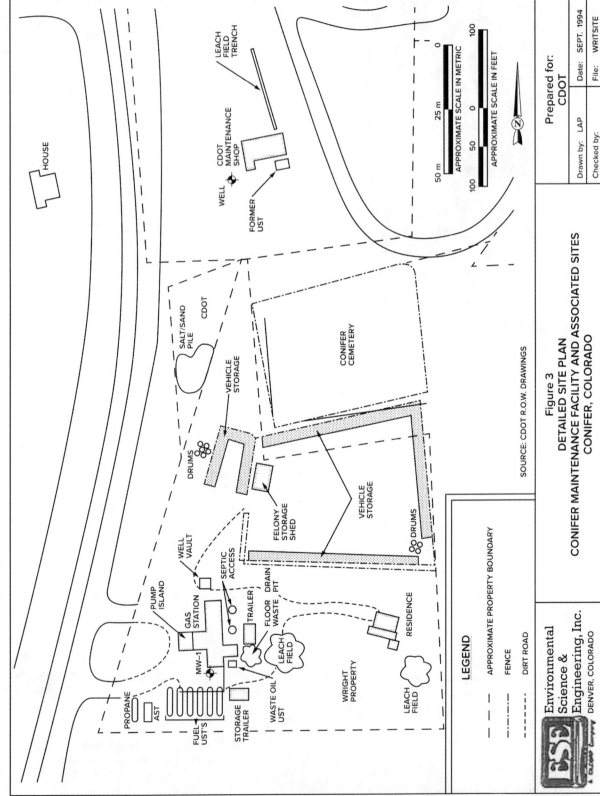

FIG. 17.4 Case study on unexpected results.

additional Phase II activities that may achieve the stated objectives unless the user redefines the objectives so that they can be met with the data available. These situations can present obstacles that can be very challenging for the consultant and user.

At times, the Phase II results indicate alternative sources of contamination. For this reason, the results of the Phase II investigations should be interpreted for indications of other sources of target analytes, or instances of higher concentrations in the environmental media, that potentially exist at the site in some locations. Some of this information may be potentially relevant to the objectives of the assessment, but it may not be specifically assessed in this assessment.

Another potential outcome is that the target analytes can occur naturally and the consultant must consider whether the detected target analytes are naturally occurring (i.e., at background concentrations), reflect diffuse anthropogenic contamination, or are present as a result of a release.

The case study in **Fig. 17.4** illustrates how unexpected Phase II ESA results can uncover new recognized environmental conditions (RECs) and result in unexpected regulatory action. It illustrates several interesting aspects of Phase II ESAs. The Phase II ESA was embarked upon to investigate the sources of contamination and to determine the responsible parties for the cleanup. The owner filed a lawsuit against what he considered a deep-pocket client. The defendant was willing to settle for a substantial amount of money. Considering the historical uses associated with his own property, the owner should have been aware of the potential for having to clean up the contamination caused by his leaking underground storage tanks.

Considering the potential for contamination, the plaintiff's attorney could have advised the client to take the settlement money offered. This could have averted the Phase II ESA being conducted by the defendant. If the attorney was not aware of the RECs and potential contamination of the subject property, then he could have hired an EP to assist in the evaluation of the contamination issues of the case. The settlement monies could have been used to offset the cost of the cleanup.

17.4 Summary

In this chapter, we learned about the scientific approach to evaluation of data to enable the consultants to interpret the results of exploration and sampling activities of Phase II ESAs to form sound, scientifically based opinions. We described the distinction between the QA and QC functions and described the individual validations steps that the consultants must complete during the evaluation process before interpreting the data. We discussed and gave examples of unexpected results and their potential implications on the consultant's ability to meet the goals of the Phase II assessments.

References

[1] ASTM Standard **E1903**, Standard Practice for Environmental Site Assessments: Phase II Environmental Site Assessment Process, *Annual Book of ASTM Standards*, ASTM International, West Conshohocken, PA, 2011.

[2] Technical & Professional Training Phase II Environmental Assessment Process, ASTM International, West Conshohocken, PA, 1999.

Chapter 18 | Phase II ESA Report

18.1 Introduction

When the **E1903** Standard [1] changed from a standard guide to a standard practice, a written report became a required element of the process. Preparation of the report may also be required by the contractual agreement between the user and the consultant. The **E1903** Standard requires a written report stating the objectives of the assessment, describing the work performed, explaining the rationale followed, and documenting the information and data acquired.

18.2 Required Elements

To comply with the standard, the written report should have three general characteristics: (1) good technical writing, (2) accurate and complete presentation of the results and conclusions, and (3) all of the supporting components of a scientific report. To meet these general characteristics, the standard set a requirement for major components listed in **Table 18.1**.

In addition to the requirements listed in **Table 18.1**, the standard also requires a specific language to be included in the report. The report must include the following statement: "We have performed a Phase II environmental site assessment at the property at (address) in conformance with the scope and limitations of ASTM Practice **E1903-11** and for the following objectives: [list 'statement of objectives' developed pursuant to Section 5.1]."

18.3 Report Format Options

Phase II Environmental Site Assessment (ESA) reports can be relatively brief or very large and complex documents. The task group members developing the standard foresaw the need for additional guidance for the consultants and prepared two sample report formats in the nonmandatory appendix to the standard. The short report typically presented in a letter format is called Option A and is shown in **Fig. 18.1**.

Option B of the report format is considerably longer, as shown in **Fig. 18.2**.

Although the **E1903** standard requires a written report, there are many legitimate reasons why the user may want the consultant not to prepare a report. The obvious one is a situation in which the user does not need a report. An example would be if the user wanted to confirm contamination at the site to decide whether to perform a site characterization. If the Phase II ESA confirms the contamination, then the user will then go directly into the more complex site characterization ESA and does not need a Phase II ESA report. Under the new **E1902** standard, this could be simply handled as an iteration of the assessment and the report is simply prepared at the conclusion of the process.

Another reason why the user may not want a report is in a situation when a finding discovered during the course of the Phase II investigation is sufficient to terminate the transaction. Spending additional efforts and money to complete the work and prepare the report would constitute a waste of time and money. Although the **E1903** Standard requires the consultant to prepare the report, the user typically has a contractual clause that allows them to terminate the assessment at any time during the process. In this situation, the user can terminate the process and not be obligated to the consultant for the preparation of the report.

An important consideration is the cost associated with preparation of the reports. Preparation of a Phase II ESA report can be a time-consuming and relatively expensive undertaking. Unlike the Phase I ESA report, which many professionals have automated by using the boilerplate, Phase II ESAs are not well suited to automation. Each report is unique and requires careful attention to detail and site-specific characteristics associated with the questions under study. Preparation of the report, when compared with other Phase II ESA activities, can take up a significant portion of the project. If given the option, the user may choose to use the generated raw data combined with verbal communications and forgo the tedious and expensive task of a report preparation. A manner in which this can be accomplished is to have a strong contractual agreement with the consultant that specifies that the user can terminate the process at any time. It is up to the user to terminate the process before the report is written. The process will not meet the requirements of the standard, but if the users have sufficient information to make a business decision, then compliance with the standard may be irrelevant. It is a good practice to discuss the standard requirements with the user and the cost of report preparation.

A well written Phase II ESA report provides the consultants with a marketing opportunity to demonstrate value, quality, and professionalism. Conversely, a poorly written report can make

TABLE 18.1   Required Major Components of a Phase II ESA Report

Required elements of a Phase II ESA written report

a) An introduction stating the objective (i.e., the question to be addressed) and including a verbatim statement of the final "Statement of Objectives" for the assessment;
b) A summary of relevant background information sufficient to explain and support the approach to the problem;
c) A description of the work performed and the rationale for it;
d) A description of the methods used;
e) A presentation of the information and data acquired;
f) Evaluation of the information and data;
g) Interpretation of the results in relation to the objective(s) and question(s) to be answered and the conceptual site model for the assessment; and
h) The signature of the Phase II assessor, together with any professional seal, license type, and license number where required by the local jurisdiction. Tables, figures, and appendices are typically included and should be used as appropriate to provide a clear and complete picture of the assessment.

valuable, quality information appear shoddy and unprofessional. Given the opportunity to submit a report at the conclusion of the Phase II ESA, the environmental professionals (EPs) should spend sufficient time and effort to demonstrate value and competence.

Preparation of a Phase II ESA report also allows the EP the opportunity to finish the process and organize the entire work product into a clear, documented, and easy-to-follow package.

18.4   Introductory Components

According to **E1903**, the intent of the transmittal or cover letter is to document the date of a report's delivery and identify the intended recipients of the report. It also may contain other important information, including a description of the report as draft or final and a designation that the report is confidential or subject to attorney work product privilege or both. The transmittal letter also may identify the Phase II assessor(s) responsible for the work.

The title page contains the title of the report, the subject property name and address, the preparer of the report, the user for whom the report was prepared, and the date of the report. It may also include a statement identifying the document as subject to certain legal privileges, such as confidentiality and attorney work product privilege.

The executive summary is a concise overview of the findings and only includes materials discussed in the body of the report. Inclusion of an executive summary is optional. A well-conceived executive summary alerts the reader to the important aspects of the report. Many consultants believe that the main reason why some users like executive summaries is that once they have read the summary, they feel no need to read further. From a practical standpoint, if the consultants want to convey pertinent information to the user, then they should include it in the executive summary. It should not simply be a copy of the conclusion section of the report. As a summary, it should generally be briefer than the conclusion section and should not contain material that is not addressed in the main text or body of the report.

18.5   Body of the Report

The introduction section identifies the purpose and scope of the Phase II ESA, the terms of the contract, conditions, and limitations. On the basis of the requirements of **E1903**, the introduction must at a minimum state the purpose and objective(s) of the Phase II ESA as set forth in the "Statement of Objectives," including any conditions or limitations on the assessment that affect the ability of the assessment to achieve the stated objectives.

E1903 also suggests identifying items not included within the scope of the Phase II ESA that might have been expected to be included (e.g., likely release areas that were not addressed and the reasons for not addressing them). This can be a difficult task because it requires the consultant to determine what would and would not be expected and it could potentially lead to exhaustive listings and explanations of nonscope items. The consultant should limit the introduction to items significant and useful to the users and the objectives of the assessment.

FIG. 18.1   Short report format from Appendix X3 of **E1903**.

X3.1   *Report Option A*—A written report compliant with this practice may be prepared in a format including the following components:

X3.1.1   Statement of Objectives, including a brief summary of background information, *user* objectives, or other factors that led to these objectives.
X3.1.2   Scope of Work, including discussion of the *conceptual model* and rationale for the assessment activities.
X3.1.3   Report on sampling activities including a description of sampling locations and their relationship to the property.
X3.1.4   Analytical data and test results.
X3.1.5   Interpretation and evaluation of data and test results in light of the *conceptual model*, leading to conclusions in relation to the objectives and questions to be answered.
X3.1.6   The written report may refer to figures, tables, and diagrams to present details of assessment activities, sampling locations, and analytical data and test results.
X3.1.7   The signature of the *Phase II assessor*, together with any professional seal, license type, and license number, where required by the local jurisdiction.
X3.1.8   Table of Contents. Report Option A is typically presented in letter format. As such, no table of contents is appropriate.

FIG. 18.2 Option B long report format.

X3.2 *Report Option B*—A written report compliant with this practice may be prepared in a format including the following components:

X3.2.1 *Title and Identification*—The identifying components of a *Phase II ESA* usually include a transmittal letter, title page, and table of contents.

X3.2.1.1 *Signatures*—The report should include the signature of the assessor, together with any professional seal, license type, and license number where required by the local jurisdiction. Signatures, along with typed names and titles may appear in the transmittal letter, on the cover page, or on a signature page at the end of the main text of the report. The signature should not be transmitted independent of the *Phase II ESA* report.

X3.2.1.2 *Transmittal Letter*—A transmittal letter documents the date of a report's delivery and identifies the intended recipients of the report. It also may contain other important information including a description of the report as draft or final and a designation that the report is confidential or subject to attorney work product privilege, or both. The transmittal letter also may identify the *Phase II assessor(s)* responsible for the work.

X3.2.1.3 *Title Page*—The title or cover page identifies the following items: the title of the report, the subject property name and address, the preparer of the report, the *user* for whom the report was prepared, and the date of the report. It may also include a statement identifying the document as subject to certain legal privileges, such as confidentiality, and attorney work product privilege.

X3.2.1.4 *Table of Contents*—The table of contents should list at least the major sections of the report, and identify the figures, tables, and appendices by name or content.

X3.2.2 *Executive Summary*—Inclusion of an executive summary is optional. It should provide a concise overview of the objectives and findings of the *ESA* and should not simply be a copy of the conclusion section of the report. As a summary, it should generally be briefer than the conclusion section and should not contain material not addressed in the main text or body of the report.

X3.2.3 *Main Text*—The main text, or body, of the report generally contains an introduction, a discussion of background information, a description of work performed during the *Phase II ESA* and the rationale behind it, descriptions of methods used, a presentation of the information and data acquired, an evaluation of the information and data, and interpretation of the results and conclusions drawn from them.

X3.2.4 *Introduction*—The introduction must at a minimum state the purpose and objective(s) of the *Phase II ESA* as set forth in the "Statement of Objectives," including any conditions or limitations on the assessment that affect the ability of the assessment to achieve the stated objectives. The introduction also should identify items not included within the scope of the *Phase II ESA* that might have been expected to be included, for example, *likely release areas* that were not addressed and the reasons for not addressing them.

X3.2.4.1 The introduction should reference this Practice **E1903** and any other applicable standards to which the work was performed, contracts/agreements between the *Phase II assessor* and *user*, regulatory requirements applicable to the work, and the general nature of the work performed (e.g., a *Phase II ESA* involving soil and *groundwater* sampling and *chemical testing*).

X3.2.5 *Background Information*—This section should include a general description of the site, its features and physical setting, a summary of site history and use and the use of adjacent properties, and a summary of the findings of previous environmental site assessments including any relevant *Phase I ESA*. Other *ESA* reports should be referenced if they provided information germane to the *Phase II ESA*.

X3.2.5.1 Pre-existing information and data that led to the *Phase II ESA* should be summarized. More detailed descriptions are to be presented later, in the discussion of the *conceptual model* and rationale for the work undertaken. If the *Phase II ESA* is based on earlier environmental site assessments, the earlier *ESAs* should be precisely identified. It is often useful (but not mandatory) to include prior *ESA* reports in an appendix.

X3.2.6 *Work Performed and Rationale*—The investigation, sampling, and testing that was accomplished should be described. The rationale for the work should be presented, including by discussing the *conceptual model* (7.4) and sampling and analysis plan (7.5) that were developed for, and refined during, the work. Any agency or facility records review, historical document review, interviews, or site reconnaissance that were conducted by the *Phase II assessor* to supplement available *Phase I ESA* information that aided in the identification of areas for investigation should be discussed. The nature and location of each area investigated should be described. The exploration and sampling locations, and samples that were field screened or tested, should be identified. The exploration, sampling, and *field screening* methods followed, the *target analytes* sought by *chemical testing*, and analytical methods, should be presented. Deviations from the intended sampling and analysis plan or from standard methodologies should be explained.

X3.2.6.1 Methods used should be identified by reference to standard methods where applicable, such as ASTM standards, state protocols, or *EPA* methodologies. Other methods used (for example, project-specific specialized methods, or standard operating procedures for a particular *Phase II Assessor*) should be described in sufficient detail that would allow another *Phase II assessor* to reproduce the work. Where appropriate, the details of specialized methods and procedures used can be included in an appendix.

(Continued)

FIG. 18.2 Option B long report format. (*Continued*)

X3.2.7 *Presentation and Evaluation of Results*—Results of the *Phase II ESA* should be organized and presented in a manner that will aid the reader in understanding the discussion and interpretation of the results in the report section to follow. The results should be presented in relation to the objective of the assessment and the question(s) addressed. Data can be tabulated and presented on figures as appropriate to aid the reader's review and understanding. The inclusion of laboratory reports of results in an appendix shall be referenced.

X3.2.7.1 Information regarding actual subsurface physical, geologic, and hydrogeologic features of the site setting that was revealed in the investigation should be described. The actual physical features should be reconciled with the assumptions of the *conceptual model* and sampling plan to validate the physical components of the *conceptual model*.

X3.2.7.2 The results should be organized and presented to assist the reader's understanding of the locations, types, and levels of *target analytes* that were encountered. The presentation may be broken down to individual area assessed. The data may be further organized with respect to *environmental media* within an area, e.g., soil, *groundwater, surface water*, soil vapor, etc. For most objectives, the data should be organized for easy comparison to regulatory criteria. For example, if the purpose of the *Phase II ESA* was to identify conditions that might cause a regulatory response, then the data should be presented along with pertinent regulatory criteria for easy comparison. Organizational divisions also may be by target analyte type (for example, organic versus inorganic *target analytes*), or other factors (e.g., the *presence* of *target analytes* in shallow soils subject to direct human exposure criteria versus impacts to deeper soils subject only to pollutant mobilization criteria), depending on the specific problems or questions addressed. Other information that should generally be presented includes: locations of sampling points in relation to potential sources and *migration pathways* (e.g., the position of a monitoring well relative to the *target analyte* source and the direction of *groundwater flow* from the source); and, the depths to pertinent horizons (e.g., the *water table*, the fill/natural soil contact, or the overburden/bedrock contact).

X3.2.7.3 The discussion of results should include the validation of the *conceptual model*, particularly confirmation that appropriate *environmental media* were sampled from optimal or other appropriate locations, and were tested for appropriate *target analytes*, in accordance with the sampling and analysis plan and as required to meet the objective of the assessment. The *substances* detected should be reconciled with the specific *target analytes* that were hypothesized to be *present*. Deviations from the sampling and analysis plan (e.g., intended samples could not be collected) must be explained so the implications of the deviations can be discussed later in the report along with interpretation and conclusions.

X3.2.8 *Interpretation and Conclusions*—Interpretation of results, as performed under Section 8, should be discussed in the report in the terms of scientific logic (i.e., statement of question, hypothesis, test of hypothesis, validation of assumptions, conclusion) in relation to the "Statement of Objectives." For each area assessed, restate the question that was to be addressed and summarize the significant attributes of the final *conceptual model* for the area (i.e., which features of the *release* or likely *release* and site setting dictated target analyte behavior, fate, and transport). Then, note the investigation tasks that were undertaken to evaluate the conditions envisioned by the *conceptual model*. Explain how the resulting data and information confirmed (or refuted) the *conceptual model*, and present the conclusions that can be drawn from the results.

X3.2.8.1 This report section should contain a summary description of *likely release areas* and areas of *presence* or likely *presence* addressed (and those relevant to the objectives of the assessment that were not addressed, if any), and the information and data generated from the investigations (including the geologic and hydrogeologic conditions encountered).

X3.2.8.2 The information and data pertinent to each area assessed should be discussed as they bear on the objective(s) of the *Phase II ESA*. Opinions and conclusions should be stated in terms corresponding to the objectives.

X3.2.8.3 The conclusion section shall clearly state whether the objectives listed in the Statement of Objectives were or were not accomplished. The conclusions must explain any deviations from planned assessment activities or limitations on what was accomplished, including the effect of such deviations or limitations on the degree of confidence or completeness of the conclusions.

X3.2.9 *Recommendations*—Recommendations for further work may be included, at the option of the *user*, consistent with the objective(s) of the *Phase II ESA*.

X3.3 *Supporting Materials*—References and sources of information should be listed in sufficient detail to allow another *Phase II assessor* to ascertain or reproduce documents and information critical to the assessment. The report should contain tables, figures, and appendices as necessary or appropriate to explain and support the main text of the report.

X3.3.1 *Tables and Figures*—Tables and figures may be used as tools for summary presentation of data. Guidelines for data organization and presentation in X3.2.7.2 apply to tables and figures as well as to text.

X3.3.1.1 Tables summarizing analytical data should also show any numerical criteria pertinent to the objectives of the assessment.

X3.3.1.2 Figures that show areas assessed, sampling points, and locations of surface and subsurface features affecting the *presence* and distribution of *target analyte* (e.g., *groundwater flow paths*) can significantly aid the reader's understanding. Drawings should be dated. Site plans should include a north arrow, a scale as appropriate to the level of accuracy of the drawing (or notation that the figure is not to scale), a legend, a title, and other appropriate identification. If figures are based on the work of others, the source and its date should be referenced.

(*Continued*)

FIG. 18.2 Option B long report format. (*Continued*)

> X3.3.2 *Appendices*—When generated during completion of the assessment, boring logs and laboratory reports, including the data validation package, should be included as appendices to the report. Other materials may be appropriate for inclusion in appendices to a *Phase II ESA* report, and should be included to the extent they help explain and document the assessment process: prior assessment reports; documents critical to the assessment but not generally available to other investigators; photographs; subsurface exploration logs; and laboratory quality control information and sample *chain of custody* forms.

The background information section of the report lists all of the pertinent information provided to the consultant in preparation and development of the scope of the assessment. This section should include a general description of the site, its features and physical setting, and a summary of site history and use and the use of adjacent properties. The background section may also include any relevant findings of previous environmental site assessments. In situations in which the historical assessments and data are critical to this assessment, many consultants include these documents in the appendices to the Phase II ESA report.

The Phase II ESA activities section of the report should accurately document the work that was performed and the rationale for the activities and methods. The rationale is derived from the discussion of the conceptual model and sampling and analysis plan that were developed for, and refined during, the assessment. This section can be used to discuss any agency or facility records review, historical document review, interviews, or site reconnaissance that were conducted by the consultant to supplement available ESA information that aided in the identification of areas for investigation.

The consultant needs to describe the nature and location of each area investigated in the assessment. The report should identify any exploration and sampling locations and samples that were field screened or tested. The exploration, sampling, and field screening methods followed; the target analytes sought by chemical testing; and analytical methods should be presented. If there were any deviations from the intended sampling and analysis plan or from standard methodologies, then they should be noted and explained in the report.

The bulk of the documentation relating to quality assurance/quality control activities and methodologies used may be presented by reference and included in appendices. This enables the peripheral information to be separated from the main body of the report and provides for clearer presentation of significant issues.

The evaluation and presentation of the results section of the report should clearly identify and separate facts, findings, assumptions, and opinions. The consultant should organize and present the results of the Phase II ESA in a manner that will enable the user to understand the points of the discussion and the interpretation of the results. The results should be presented in relation to the objectives of the assessment that are contained in the questions that were addressed. Data can be tabulated and presented in figures as appropriate to aid the user's review and understanding.

Interpretation of results should be discussed in the report in the terms of scientific logic, stating the question, hypothesis, test of hypothesis, validation of assumptions, and conclusion in relation to the Statement of Objectives.

The information and data pertinent to each area assessed should be discussed as they bear on the objective(s) of the Phase II ESA. Opinions and conclusions should be stated in terms corresponding to the objectives.

The conclusion section shall clearly state whether the objectives listed in the Statement of Objectives were or were not accomplished. The conclusions must explain any deviations from planned assessment activities or limitations on what was accomplished, including the effect of such deviations or limitations on the degree of confidence or completeness of the conclusions.

Recommendations for further work may be included, at the option of the user, consistent with the objective(s) of the Phase II ESA.

The report should include the signature of the assessor, together with any professional seal, license type, and license number where required by the local jurisdiction. Signatures, along with typed names and titles, may appear in the transmittal letter, on the cover page, or on a signature page at the end of the main text of the report. The signature should not be transmitted independent of the Phase II ESA report.

Many consultants also include a technical review and signature by other professional members of the firm. The technical content review may be performed by a person not necessarily involved in the Phase II ESA. The review provides an additional level of quality control and may note errors or inconsistencies not readily apparent to someone closely involved with the project.

The consultant should provide a list of references and sources of information in sufficient detail to allow another Phase II assessor to ascertain or reproduce supporting materials and information critical to the assessment. The report should contain tables, figures, and appendices as necessary or appropriate to explain and support the main text of the report.

18.6 Appendices

Phase II ESA reports often include many items that were used to develop the scope of the work, conclusions, and opinions. The bulk of the material does not need to be presented in the body of the report. Referencing the information and providing it in appendices makes the main body of the report less bulky and easier to read. The consultant should consider who is going to read the report.

Although scientific information is being presented, the target audience may not have extensive scientific background, and some of the information may be confusing.

Appendices will typically include boring logs and laboratory reports, including the data validation package. Other materials may be appropriate for inclusion in appendices to a Phase II ESA report and should be included to the extent that they help explain and document the assessment process, such as prior assessment reports, documents critical to the assessment but not generally available to other investigators, photographs, subsurface exploration logs, and laboratory quality control information and samples.

The information in the appendices should be logically organized. If numerous appendices are being presented, then an index to appendices should be provided. It is also useful to clearly separate the appendices with visual aids such as colored pages or tabs to enable the reader to quickly access the section of interest. Providing curriculum vitae and a company's scope of services statements as one of the appendices provides another soft marketing opportunity for the consultants.

18.7 Summary

In this chapter, it was noted that the 2011 **E1903** Standard now requires a written report. We reviewed the optional format that the standard provides and discussed various aspects related to the individual components of the report.

References

[1] ASTM Standard **E1903**, Standard Practice for Environmental Site Assessments: Phase II Environmental Site Assessment Process, *Annual Book of ASTM Standards*, ASTM International, West Conshohocken, PA, 2011.

Chapter 19 | Dealing with Nonscope Issues in Phase II Assessments

19.1 Introduction

When the **E1903** Standard [1] changed from standard guide to standard practice, the scope of the types of investigations that could be included became significantly broader. Many of the issues that were previously beyond the scope of the Phase II ESA, such as lead-based paint, vapor intrusion, issues beyond Comprehensive Environmental Response, Compensation, and Liability Act (CERCLA)/Landowner Liability Protections (LLPs), and other business decision driven investigations, can now be a part of the Phase II Environmental Site Assessment (ESA) process, and the standard can be applied to these investigation. This was one of the major driving forces behind the 2011 revision of the Phase II standard.

There are some issues that are beyond the scope of the standard, and they are discussed in this chapter.

19.2 Issue-Specific Standards and Regulations

For any scientifically based investigation, it is appropriate to review and use any applicable standards or regulations if they are available. In some instances, such as asbestos, there may be specific standards that need to be followed, and it may not be appropriate for the consultant to investigate these using the Phase II ESA standard. The consultant needs to be aware of local regulations and design the investigation to meet the applicable standards.

19.3 Contracting And Subcontracting

Contracting and subcontracting issues remain an out-of-scope item that the consultants need to consider.

Contracting issues were discussed in **Chapter 5**. They are invariably a part of every Phase II ESA. During the Phase I ESA, the contract can simply state that the consultant agrees to perform the assessment in accordance with the scope of **E1527** [2]. The tasks associated with Phase I ESAs are sufficiently defined and specified in **E1527**. By contrast, the **E1903** scope is very broad, and the Phase II standard only lists contracting considerations in the nonmandatory appendix. Therefore, the user and the consultant must develop a mutually agreeable contract.

The subcontracting portion(s) of the work associated with Phase II ESA activities to other companies or individuals adds the inherent risk of being sued along with any of the subcontractors. It is common in Phase II investigations to subcontract various portions of the investigation, such as drill and direct-push rigs (**Fig. 19.1**). Lawsuits generally try to include as many parties as possible. As demonstrated in **Chapter 15**, the consultant can be held partially responsible for errors or wrongdoing by any of the subcontractors.

It is noteworthy that many subcontractors such as in the unit depicted do not engage in visible advertising, such as on the side of the equipment. This is sometimes called for in contracts because several clients may be sensitive to advertising to casual observers that any environmental investigation is going on at their property. However, some states require that the name and company name be displayed, especially for services such as water drilling, waste hauling, etc.

19.4 Safety and Health

In **Chapter 7**, we discussed the safety and health issues that may be associated with Phase I and Phase II ESAs. They are considered beyond the scope of **E1903**. Many of these issues are specific to site and transaction and must be taken into consideration by the consultants. The consultants that conduct environmental site investigations are responsible for their own safety and often also have responsibilities with respect to the subcontractor and site occupants' or invitees' safety. Training and experience in health and safety issues appropriately applied during Phase II ESAs help to ensure safe and successful completion of projects.

19.5 Recommendations

Recommendations are not a required element in the **E1903** Phase II ESA. The standard specifically states that recommendations are subject to contractual agreement. The consultant should communicate closely with the user and follow the contract.

19.6 Business Judgments

What the user decides to do with the information and the results generated from the Phase II ESA is up to the user and is not

FIG. 19.1   GeoProbe services are frequently subcontracted.

Source: Courtesy of JAEE Environmental, Davie, FL (photograph by Zdenek Hejzlar).

discussed in the **E1903** Standard. Business judgment often extends beyond the scope of any work that the consultants provide for the user, although the consultants input the result, and the report can play a significant role in the judgments.

19.7   Legal Risks

E1903 does not address many of the legal risks that may be associated with performing Phase II assessments. Appendix X1 of the standard presents several legal considerations that could play a significant role in the process. The consultant's business judgment and risk tolerance also play an important role in dealing with these issues. By the act of hiring the consultants to perform Phase II ESAs, the users are spreading some of their liabilities to the consultant. This must be taken into consideration, and through contract/insurance and quality of work, the consultants need to adequately minimize the liability potential.

19.8   Summary

This chapter reviewed and summarized some of the many issues that are an integral part of Phase II ESAs and are not specifically addressed by or included in the scope of the **E1903** Standard. Previous chapters provided several specific examples that demonstrated practical applications of the concepts outlined in this chapter.

References

[1] ASTM Standard **E1903**, Standard Practice for Environmental Site Assessments: Phase II Environmental Site Assessment Process, *Annual Book of ASTM Standards,* ASTM International, West Conshohocken, PA, 2011.

[2] ASTM Standard **E1527**, Standard Practice for Environmental Site Assessments: Phase I Environmental Site Assessment Process, *Annual Book of ASTM Standards,* ASTM International, West Conshohocken, PA, 2013.

Appendix: Forms

Building Permit Review

Reviewer: _____
Date of Review: _____

Project Number: _____ Project Name: _____
Agency: _____
Property Address: _____
APN: _____ Zoning: _____ Lot Size: _____
Legal Description: _____

Permit Date	Type of Permit	Notes and Comments (owner name, bldg. size, env. concerns, etc.)
	☐ Building ☐ Alteration, Repair, Demolition ☐ Plumbing ☐ Electrical ☐ Sewer ☐ Mechanical ☐ Other:	
	☐ Building ☐ Alteration, Repair, Demolition ☐ Plumbing ☐ Electrical ☐ Sewer ☐ Mechanical ☐ Other:	
	☐ Building ☐ Alteration, Repair, Demolition ☐ Plumbing ☐ Electrical ☐ Sewer ☐ Mechanical ☐ Other:	
	☐ Building ☐ Alteration, Repair, Demolition ☐ Plumbing ☐ Electrical ☐ Sewer ☐ Mechanical ☐ Other:	
	☐ Building ☐ Alteration, Repair, Demolition ☐ Plumbing ☐ Electrical ☐ Sewer ☐ Mechanical ☐ Other:	
	☐ Building ☐ Alteration, Repair, Demolition ☐ Plumbing ☐ Electrical ☐ Sewer ☐ Mechanical ☐ Other:	
	☐ Building ☐ Alteration, Repair, Demolition ☐ Plumbing ☐ Electrical ☐ Sewer ☐ Mechanical ☐ Other:	

This form may be copied only if the following attribution is retained: Source: *Technical Aspects of Phase I/II Environmental Site Assessments, 3rd Ed.* © ASTM International. All rights reserved. Used with permission. There is no further fee or additional permission required.

Property Reconnaissance[1]

Date of Reconnaissance: _____

Assessor: _____

Project Number: _____ Project Name: _____

Property Address/Location: _____

Property Use: _____

Property/Lot Size: _____

Surface Topography: _____

Adjacent Properties: N: _____

NE: _____

NW: _____

S: _____

SE: _____

SW: _____

E: _____

W: _____

General Vicinity: _____

Photographs Taken: ☐ Yes ☐ No   Prop. Map Drawn: ☐ Yes ☐ No

Building Description

No. of Buildings: _____

No. of Stories per Building: _____

Age of Building(s): _____

Areas/Units Accessed: _____

Areas/Units Not Accessed: _____

---

[1] ASTM **E1527-13**, Standard Practice for Environmental Site Assessments: Phase I Environmental Site Assessment Process, ASTM International, West Conshohocken, PA, 2013, www.astm.org.

Unoccupied Occupant Spaces: _____
_____

Construction Type:	☐ Wood Frame	☐ Steel Frame	☐ Cast-in-Place	☐ Masonry	☐ Concrete Tilt-Up	☐ Other
Exterior:	☐ Concrete	☐ Block	☐ Wood	☐ Glass	☐ Stucco	☐ Brick ☐ Other

Floors/Decking:	☐ Wood	☐ Concrete	☐ Metal	☐ Other		
Interior Walls:	☐ Concrete	☐ Block	☐ Wood	☐ Drywall	☐ Stucco	☐ Plaster ☐ Other

Ceilings:	☐ Concrete	☐ Tile	☐ Wood	☐ Drywall	☐ Sprayed Acoustical	☐ Plaster ☐ Other

Roof:	☐ Gravel	☐ Built-Up	☐ Shingle	☐ Tile	☐ Hypolon	☐ Foam ☐ Other

HVAC:	☐ Central	☐ Window Mount	☐ Wall Mount	☐ Floor Vents	☐ Roof Mount	☐ Other

Insulation:	☐ Fiberglass	☐ Blown-In	☐ Cellulose	☐ Other

Item or Condition	Interior	Exterior	Comments
Hazardous Substances and Petroleum Products:	☐	☐	
Storage Tanks and Related Equipment:	☐	☐	
Odors:	☐	☐	
Standing Surface Water or Other Pools of Liquid:	☐	☐	
Drums and Other Containers of Hazardous Substances, Petroleum Products, or Other Unidentified Contents:	☐	☐	
Transformers or Equipment Containing Polychlorinated Biphenyls (PCBs):	☐	☐	
Stains and Corrosion:	☐	☐	
Drains and Sumps:	☐	☐	
Pits, Ponds, or Lagoons:	☐	☐	
Stained Soil or Pavement:	☐	☐	
Stressed Vegetation (Other Than from Insufficient Water):	☐	☐	
Evidence of Mounds, Depressions, or Filled or Graded Areas Suggesting Trash or Other Solid Waste Disposal:	☐	☐	
Waste Water or Any Discharge (Including Storm Water) into a Drain, Ditch, or Stream on or Adjacent to the Property:	☐	☐	
Wells (active, inactive, or abandoned):	☐	☐	
Septic Systems or Cesspools:	☐	☐	
Prior Structures:	☐	☐	
Roads, Tracks, Railroad Tracks or Spurs, Parking Facilities:	☐	☐	

Additional Reconnaissance Comments:

Sewer or Septic System: _____
Potable Water Supplier: _____
Utility Provider – Gas: _____
Utility Provider – Electricity: _____
Provider – Solid Waste: _____

On-site Interviews: _____

Additional Information: _____

Page ____ of ____

ASTM 1527-13 APPENDIX X3—User Questionnaire[1]

In order to qualify for one of the Landowner Liability Protections (LLPs)[2] offered by the Small Business Liability Relief and Brownfields Revitalization Act of 2001 (the "Brownfields Amendments"),[3] the user must conduct the following inquiries required by 40 CFR 312.25, 312.28, 312.29, 312.30, and 312.31. These inquiries must also be conducted by EPA Brownfield Assessment and Characterization grantees. The user should provide the following information to the environmental professional. Failure to conduct these inquiries could result in a determination that "all appropriate inquiries" is not complete.

1. **Environmental liens that are filed or recorded against the *property* (40 CFR 312.25).**
   Did a search of *recorded land title* records (or judicial records where appropriate, see Note 1 below) identify any environmental liens filed or recorded against the *property* under federal, tribal, state or local law?
   Note 1—In certain jurisdictions, federal, tribal, state, or local statutes, or regulations specify that environmental liens and AULs be filed in judicial records rather than in land title records. In such cases judicial records must be searched for environmental liens and AULs.

2. **Activity and land use limitations that are in place on the *property* or that have that been filed or recorded against the *property* (40 CFR 312.26(a)(1)(v) and (vi)).**
   Did a search of recorded land *title records* (or judicial records where appropriate, see Note 1 above) identify any AULs, such as *engineering controls*, land use restrictions or *institutional controls* that are in place at the *property* and/or have been filed or recorded against the *property* under federal, tribal, state or local law?

3. **Specialized knowledge or experience of the person seeking to qualify for the LLP (40 CFR 312.28).**
   Do you have any specialized knowledge or experience related to the *property* or nearby properties? For example, are you involved in the same line of business as the current or former *occupants* of the *property* or an *adjoining property* so that you would have specialized knowledge of the chemicals and processes used by this type of business?

4. **Relationship of the purchase price to the fair market value of the *property* if it were not contaminated (40 CFR 312.29).**
   Does the purchase price being paid for this *property* reasonably reflect the fair market value of the property? If you conclude that there is a difference, have you considered whether the lower purchase price is because contamination is known or believed to be present at the property?

5. **Commonly known or *reasonably ascertainable* information about the *property* (40 CFR 312.30).**
   Are you aware of commonly known or *reasonably ascertainable* information about the *property* that would help the *environmental professional* to identify conditions indicative of releases or threatened releases? For example,

   a) Do you know the past uses of the *property*?

   b) Do you know of specific chemicals that are present or once were present at the *property*?

   c) Do you know of spills or other chemical releases that have taken place at the *property*?

   d) Do you know of any environmental cleanups that have taken place at the *property*?

---

[1] ASTM **E1527-13**, Standard Practice for Environmental Site Assessments: Phase I Environmental Site Assessment Process, ASTM International, West Conshohocken, PA, 2013, www.astm.org.
[2] Landowner Liability Protections, or LLPs, is the term used to describe the three types of potential defenses to Superfund liability in EPA's *Interim Guidance Regarding Criteria Landowners Must Meet in Order to Qualify for Bona Fide Prospective Purchaser, Contiguous Property Owner, or Innocent Landowner Limitations on CERCLA Liability ("Common Elements"* Guide) issued on March 6, 2003.
[3] P.L. 107–118.

6. **The degree of obviousness of the presence or likely presence of contamination at the *property*, and the ability to detect the contamination by appropriate investigation (40 CFR 312.31).**
   Based on your knowledge and experience, related to the *property* are there any *obvious* indicators that point to the presence or likely presence of releases at the *property*?

X3.1 In addition, certain information should be collected, if available, and provided to the *environmental professional* conducting the *Phase I Environmental Site Assessment*. This information is intended to assist the *environmental professional*, but is not necessarily required to qualify for one of the *LLPs*. The information includes:

a. The reason why the Phase I is being performed,
b. The type of *property* and type of *property* transaction, for example, sale, purchase, exchange, etc.,
c. The complete and correct address for the *property* (a map or other documentation showing *property* location and boundaries is helpful),
d. The scope of services desired for the Phase I (including whether any parties to the *property* transaction may have a required standard scope of services or whether any considerations beyond the requirements of Practice E1527 are to be considered),
e. Identification of all parties who will rely on the Phase I *report,*
f. Identification of the site contact and how the contact can be reached,
g. Any special terms and conditions which must be agreed upon by the *environmental professional*, and
h. Any other knowledge or experience with the *property* that may be pertinent to the *environmental professional* (for example, copies of any available prior *environmental site assessment reports*, documents, correspondence, etc., concerning the *property* and its environmental condition).

Name (Signature): _____

Name (Print): _____

Date: _____

This form may be copied only if the following attribution is retained: Source: *Technical Aspects of Phase I/II Environmental Site Assessments, 3rd Ed.* © ASTM International. All rights reserved. Used with permission. There is no further fee or additional permission required.

User Responsibility Statement[1]

In an attempt to obtain information on recognized environmental conditions in connection with the property, Consultant will make reasonable attempts to interview current owners and occupants of the property. The Client agrees to assist in identifying the owner or key personnel who may have knowledge of the uses and physical characteristics of the property.

Consultant understands that the Client is the only intended user of the document. If it is the intent to have other parties rely on the Phase I ESA report, they must be indicated to Consultant prior to the issuance of the report. It is important that the other relying parties be made aware of the information which follows regarding their responsibilities if they wish to satisfy the requirements for *All Appropriate Inquiry*.

The Client agrees to provide any information available to the Client or other Users relating to past and current uses of this property and to the past and/or current contamination of this property by toxic or hazardous substances. It is the Client/Users responsibility to search for reasonably ascertainable recorded land title records for environmental liens and activity use limitations. Anyone seeking defenses to CERCLA liability must take independent action, including continuing obligations if applicable, to perfect their position.

In order to facilitate the collection of useful information, Consultant may provide forms for your use relating to ASTM **1527-13** Appendix X3 *User Questionnaire* and Section 6 *User's Responsibilities*. These forms may assist you in complying with the statutory requirements of the AAI law.

The Introduction to ASTM X3 states:

> *In order to qualify for one of the Landowner Liability Protections (LLPs) offered by the Small Business Relief and Brownfields Revitalization Act of 2001 (the "Brownfields Amendments"), the user must conduct the following inquiries required by 40 CFR 312.25, 312.28, 312.29, 312.30, and 312.31. These inquiries must also be conducted by EPA Brownfield Assessment and Characterization grantees. The user should provide the following information to the environmental professional. Failure to conduct these inquiries could result in a determination that "all appropriate inquiry" is not complete.*

---

[1] ASTM **E1527-13**, Standard Practice for Environmental Site Assessments: Phase I Environmental Site Assessment Process, ASTM International, West Conshohocken, PA, 2013, www.astm.org.
This form, may be copied only if the following attribution is retained: *Source: Technical Aspects of Phase I/II Environmental Site Assessments, 3rd Ed.* © ASTM International. All rights reserved. Used with permission. There is no further fee or additional permission required.

Owner Interview and Helpful Documents[1]

Pursuant to our scope of services, Converse Consultants requests your knowledge as to the existence of any documents matching, or relating to, those specified below. Please indicate whether copies of the information will be provided for our review. We also request a response to site specific questions. Thank you for your cooperation.

DOCUMENTS

Document or Information	I Have Knowledge of the Information		Copies will be Provided to Converse	
	Yes	No	Yes	No
Environmental site assessment reports				
Environmental compliance audit reports				
Environmental permits, hazardous waste, generator notices, or reports				
Aboveground and underground tank and injection well registrations or related information				
Disclosure of sumps, pits, drainage systems–existence and location				
Risk assessments				
Safety Data Sheets (SDS)				
Community right-to-know plans				
Safety, preparedness and prevention plans; spill protection countermeasures and control plans; facility response plans				
Hydrogeological reports and geotechnical studies				
Notices or other correspondence from any government agency relating to past or existing environmental liens encumbering the property, or violations				
Information concerning any pending, threatened, or past litigation, or administrative proceedings relevant to hazardous substances or petroleum products in or on the property				
Notices from any governmental entity regarding any possible violation of environmental laws or possible liability relating to hazardous substances or petroleum products				
Any specialized knowledge or experience that is material to recognized environmental conditions in connection with the subject property				
Title records for environmental liens or activity and land use limitations				
Title records for environmental liens or activity and land use limitations				
Valuation reduction for environmental issues				
Environmental problems with adjacent or vicinity locations				
Building plans				
Current and historical photographs of the property				

COMMENTS:

_____
_____
_____
_____

[1] ASTM **E1527-13**, Standard Practice for Environmental Site Assessments: Phase I Environmental Site Assessment Process, ASTM International, West Conshohocken, PA, 2013, www.astm.org.

QUESTIONS

Current site use and length of ownership:

Prior site use:

Sewage disposal system/septic system on-site:

Use, treatment, storage, disposal, and generation of hazardous waste and petroleum products:

Presence of underground and aboveground storage tanks (USTs and ASTs):

Potable water supply:

Means of heating and cooling for structures on-site:

Unidentified contents of observed containers:

Oil or water wells on-site:

Any discharge of waste water onto property or adjacent property:

Environmental problems with adjacent or vicinity locations:

Other information or comments:

Name/Title of Responder _____ **Date**_____

This form may be copied only if the following attribution is retained: Source: *Technical Aspects of Phase I/II Environmental Site Assessments, 3rd Ed.* © ASTM International. All rights reserved. Used with permission. There is no further fee or additional permission required.

Table of Contents Template Discussion[1]

ASTM **1527-13** Appendix X4 provides a recommended Table of Contents and Report format as follows:

- Summary
- Introduction
- User Provided Information
- Records Review
- Site Reconnaissance
- Interviews
- Evaluation
- Non-Scope Services
- Appendices

Keep in mind the above noted table of contents is not mandatory so long as any alternate used complies with Section 12, *Evaluation and Report Preparation* of the Standard. The alternate version may have the Site Reconnaissance section described before there is a discussion of the Records Review. As noted in Section 12.1, user requirements may be considered and some users may wish to understand elements of the Site Reconnaissance prior to discussion of records detailing potential off-site impacts.

An alternate Table of Contents and Report format could be:

- Executive Summary
- Introduction
- Property Description
- User/Owner Provided Information and Responsibilities
- Site Reconnaissance
- Records Review
- Interviews
- Findings
- Opinion
- Conclusions
- Non-Scope Services
- References
- Appendices

---

[1] ASTM **E1527-13**, Standard Practice for Environmental Site Assessments: Phase I Environmental Site Assessment Process, ASTM International, West Conshohocken, PA, 2013, www.astm.org.
This form may be copied only if the following attribution is retained: *Source: Technical Aspects of Phase I/II Environmental Site Assessments, 3rd Ed.* © ASTM International. All rights reserved. Used with permission. There is no further fee or additional permission required.

Index

A

AAI. *See* All Appropriate Inquiries Final Rule
abandoned properties, 58–59 (figure)
    interviewing owners or occupants of, 57–58
    safety and health, 51, 54
aboveground storage tanks, 20 (figure), 43 (figure), 51, 97 (figure)
Activity and Use Limitations (AULs), 25, 62, 139
actual knowledge, defined, 61
acute exposure, 51
additional investigations, including in Phase I ESA reports, 66, 66 (figure)
additional services, including recommendations for in Phase I ESA reports, 67 (figure), 67–68, 72
adjoining properties
    concept of, 39, 39 (figure)
    identifying uses of, 40
    mandatory record information, 33, 34 (figure)
    reviewing current and past uses of, 38–40
agency file reviews, 32–33, 34 (figure)
AHERA (Asbestos Hazard Emergency Response Act), 3
All Appropriate Inquiries (AAI) Final Rule, 2, 3, 8
    contractual considerations regarding Brownfields, 99
    criteria for in Brownfields Amendments, 22 (figure), 22–23
    data gaps, 36, 66, 66 (figure)
    interviewing neighboring property owners, 57
    objectives of, 22 (figure)
    players in Phase I ESAs, 9
    practical application of Phase I ESAs, 19–20
    practical application of Phase II ESAs, 89–90
    principles behind, 22–23
    recommendations for additional investigations, 66, 66 (figure)
    requirement and checklist, 86–87 (figure)
    responsibilities under, 25–26
    responsible charge, 28
    scope of Phase I ESAs, 11
    scope of Phase II ESAs, 77
    terminology, 15, 15 (figure)

American Society of Civil Engineers (ASCE), 18
appendices, including in reports
    Phase I ESAs, 68–69
    Phase II ESAs, 129 (figure), 129–130
approximate minimum search distances, 33 (figure), 64
AQR Color-Tec analytical method, 111, 114 (figure)
asbestos, 3, 64, 71, 85
Asbestos Hazard Emergency Response Act (AHERA), 3
ASCE (American Society of Civil Engineers), 18
assessment safety. *See* safety and health
ASTM Standard D653, Terminology Relating to Soil, Rock, and Contained Fluids, 18, 85
ASTM Standard D4220, Standard Practice for Preserving and Transporting Soil Samples, 114
ASTM Standard D4750, Standard Test Method for Determining Substrate Liquid Level in a Borehole or Monitoring Well, 112
ASTM Standard D5730, Standard Guide to Site Characterization for Environmental Purposes with Emphasis on Soil, Rock, the Vadose Zone, and Groundwater, 85
ASTM Standard D6001, Standard Guide for Direct-Push Groundwater Sampling for Environmental Site Characterization, 111
ASTM Standard E620, Standard Practice for Reporting Opinions of Technical Experts, 66
ASTM Standard E678, Standard Practice for Evaluation of Technical Data, 66
ASTM Standard E1527, Standard Practice for Environmental Site Assessments: Phase I Environmental Site Assessment Process, 7–12, 19–24
    buyers, 9
    environmental professionals (EPs), 9
    history of, 7–8
    identifying RECs, 20–21
    interviews, 57–62
    lenders, 8–9
    limitations of, 10, 11
    nonscope issues, 71–73

objectives of, 10
principles behind, 23
principles behind AAI, 22–23
purpose of, 10
records review, 31–36
regulators, 9–10
relationship to E1528, 21–22
report preparation, 63–69
responsibilities under, 25–30
rules of engagement, 24
safety and health, 51–55
scope of, 10–11
shelf life of reports, 23–24
site reconnaissance, 37–49
standardized terminology, 13–18
users not seeking LLPs, 20
voluntary use of, 19
ASTM Standard E1528, Standard Practice for Limited Environmental Due Diligence: Transaction Screen Process, 17, 18 (figure), 21 (figure), 22, 78, 85
E1527 relationship to, 21–22
Transaction Screen questionnaire, 61
ASTM Standard E1689, Guide for Developing Site Models for Contaminated Sites, 83
ASTM Standard E1903, Standard Guide for Environmental Site Assessments: Phase II Environmental Site Assessment Process, Historical Standard, 77–81, 107–117, 108 (figure)
applicability of, 89–92
conceptual model validation, 109
conceptual models, 107–108
conclusions, 109
contracts, 95–99
environmental media sampling, 111–112
field screening and analytical techniques, 109–111
limitations of, 80–81, 93–94 (figure), 93–94
nonscope considerations, 131–132
objectives of, 78–80, 90 (table), 96, 107
report preparation, 109, 125–130
results interpretation, 121–123
revisions to, 77
sample handling, 112–116
sampling, 109
sampling and testing plans, 108–109
scope of, 77–78
scope of work, 101–105
standardized terminology, 83–88
users' needs, 80, 92 (figure)
validation process, 119–121, 120 (figure)
ASTM Standard E2091, Standard Guide for Use of Activity and Use Limitations, Including Institutional and Engineering Controls, 17

ASTM Standard E2247 for Environmental Site Assessments: Phase I Environmental Site Assessment for Forestland or Rural Property, 8, 78, 85, 94
ASTM Standard E2531 Guide for Development of Conceptual Site Models and Remediation Strategies for Light Non-Aqueous Phase Liquids Released to the Subsurface, 83
AULs (Activity and Use Limitations), 25, 62, 139

B

beehives, 51, 52 (figure)
BER (Business Environmental Risk) appendix, 72
Bona Fide Prospective Purchaser (BFPP) defense, 2, 8
continuing obligations, 16
practical application of Phase II ESAs, 89, 91 (figure)
reporting obligations and confidentiality, 96 (figure)
scope of Phase II ESAs, 77
windfall liens, 25–26
Brownfields Amendments of 2002, 2–3, 7–9, 89–90
criteria for AAI, 22 (figure), 22–23
limitations of Phase I ESAs, 11
Phase II assessors (consultants), 84, 84 (figure)
practical application of Phase I ESAs, 20–21
responsibilities under AAI, 26–27
scope of Phase II ESAs, 77
Brownfields and grant program, 9–10, 23
checklist of specific requirements, 85, 85 (figure)
contractual considerations, 99
objectives of Phase II ESAs, 79 (figure), 90 (figure)
practical application of Phase II ESAs, 89–90, 90, 91 (figure)
work plans, 101
Building Permit Review form, 134
Bush, George W., 7
business environmental risk, 65, 67 (figure), 79, 79 (figure), 90–91, 91 (figure)
Business Environmental Risk (BER) appendix, 72
buyers, defined, 9

C

CAA (Clean Air Act) of 1970, 2
CERCLA. *See* Brownfields Amendments; Comprehensive Environmental Response, Compensation, and Liability Act; Landowner Liability Protections
chain of custody, 83, 115, 117 (figure)
chemical analysis error, 94
chemical storage, 21, 21 (figure). *See also* underground storage tank systems
chemical testing plans, 104
chronic exposure, 51
citrus canker decontamination stations, 51, 52 (figure)
Clean Air Act (CAA) of 1970, 2
Clean Water Act (CWA) of 1977, 2

Comprehensive Environmental Response, Compensation, and Liability Act (CERCLA; "Superfund Law")
of 1980, 2, 7, 19
    environmental liens, 25
    liens, 25, 26 (figure)
    objectives of Phase II ESAs, 79 (figure), 90 (table)
    practical application of Phase I ESAs, 19–20
    practical application of Phase II ESAs, 89–90, 90 (figure)
    scope of Phase I ESAs, 10–11
    Secured Creditor Exemption, 3, 9, 22
    Superfund Amendments and Reauthorization Act, 3, 7, 22
    terminology, 83
    work plans, 101–102
    See also Brownfields Amendments; Landowner Liability Protections
conceptual site model (CSM), 83, 84 (figure)
    development of, 107–108, 108–109 (table)
    thinking process, 78, 78 (figure)
    validating, 109
conclusions
    developing, 109
    including in Phase I ESA reports, 67, 67 (figure)
    including in Phase II ESA reports, 126 (figure), 128 (figure), 129
confidentiality issues, 26, 95–96
construction debris, 13, 14 (figure)
content of interviews, 60–61
contents of reports
    Phase I ESAs, 63
    Phase II ESAs, 126–127 (figure)
    template for, 144
Contiguous Property Owner (CPO) defense, 2, 8
contracts, 95–99
    conceptual definition of, 95
    generation of waste and exploration damage, 97
    as nonscope issue in Phase II ESAs, 131
    reporting obligations and confidentiality, 95–96, 96 (figure)
    responsibilities of users and consultants, 97–99
    subcontracting, 55, 97, 97 (figure), 131, 132 (figure)
    written communication, 96
controlled recognized environmental conditions (CRECs), 16, 17 (figure)
    defined, 16 (figure), 65, 65 (figure)
    including in reports as findings, 64
    records review, 35
CPO (Contiguous Property Owner) defense, 2, 8
creative photography, 49
CRECs. *See* controlled recognized environmental conditions
CSM. *See* conceptual site model
CWA (Clean Water Act) of 1977, 2

D

D653, Terminology Relating to Soil, Rock, and Contained Fluids, 18, 85
D4220, Standard Practice for Preserving and Transporting Soil Samples, 114
D4750, Standard Test Method for Determining Substrate Liquid Level in a Borehole or Monitoring Well, 112
D5730, Standard Guide to Site Characterization for Environmental Purposes with Emphasis on Soil, Rock, the Vadose Zone, and Groundwater, 85
D6001, Standard Guide for Direct-Push Groundwater Sampling for Environmental Site Characterization, 111
data failure, 36
data gaps, 31, 36
    defined, 66 (figure)
    including in reports, 66–67
data quality objectives, 102
decontamination procedures, 54–55
deep pocket provision, 1
demolition debris, 13, 14 (figure)
deviations, including in reports, 68
distribution of contaminants, 103, 114
document retention policies, 68
documentation
    of collected samples, 113–114, 116 (figure)
    including in Phase I ESA reports, 63
    including in Phase II ESA reports, 127–128 (figure), 129
drive-by site reconnaissance, 38 (figure)
dumping, 39, 40 (figure)

E

E620, Standard Practice for Reporting Opinions of Technical Experts, 66
E678, Standard Practice for Evaluation of Technical Data, 66
E1527 Standard. *See* ASTM Standard E1527, Standard Practice for Environmental Site Assessments: Phase I Environmental Site Assessment Process
E1528 Standard. *See* ASTM Standard E1528, Standard Practice for Limited Environmental Due Diligence: Transaction Screen Process
E1689, Guide for Developing Site Models for Contaminated Sites, 83
E1903 Standard. *See* ASTM Standard E1903, Standard Guide for Environmental Site Assessments: Phase II Environmental Site Assessment Process, Historical Standard
E2091, Standard Guide for Use of Activity and Use Limitations, Including Institutional and Engineering Controls, 17
E2247 for Environmental Site Assessments: Phase I Environmental Site Assessment for Forestland or Rural Property, 8, 78, 85, 94

E2531 Guide for Development of Conceptual Site Models and Remediation Strategies for Light Non-Aqueous Phase Liquids Released to the Subsurface, 83
Emergency Planning and Community Right-to-Know Act (EPCRA), 3
engineering controls (ECs), 16–17, 25, 26 (figure), 79, 91
environmental liens, 25, 26 (figure)
environmental professionals (EPs), 9
    chemical research, 54
    defined, 28 (figure), 84 (figure)
    evaluation of data, 63
    identifying in reports, 63
    identifying RECs, 20–21
    identifying uses of adjoining properties, 40
    including conclusions in reports, 67, 67 (figure)
    including opinions in reports, 64–66
    including recommendations in reports, 67 (figure), 67–68, 72
    including statements of in reports, 68, 68 (figure)
    information request form, 27 (figure)
    responsibilities of, 28, 84
    See also interviews; Phase II assessors; site reconnaissance
EPCRA (Emergency Planning and Community Right-to-Know Act), 3
EPs. *See* environmental professionals; interviews; Phase II assessors; site reconnaissance
exploration damage, 97, 98 (figure), 114
exposure to toxic or chemically active substances, 51
exterior observations during site reconnaissance, 43–45

F
Federal Water Pollution Act (FW-PCA) of 1972, 2
field screening and analytical techniques, 109–111, 110–115 (figure)
findings
    including in reports, 64
    opinions on, 64–66
flow-through cell and low-flow sample collection, 113, 115 (figure)
FW-PCA (Federal Water Pollution Act) of 1972, 2

G
gas chromatographs (GCs), 111, 114 (figure)
general observations during site reconnaissance, 42–43
general site setting, 38–40
government officials, interviews with, 61–62
GPS, 46 (figure)
ground-penetrating radar, 110, 110 (figure)

H
Hazardous Materials Transportation Uniform Safety Act (HMTUSA) of 1990, 3
Hazardous Solid Waste Amendments (HSWA) of 1984, 3

HAZWOPER Standard, 53, 54
health and safety (H&S) plans, 104. *See also* safety and health
helpful documents, 26–28, 27 (figure), 58, 61, 142
historical recognized environmental conditions (HRECs)
    defined, 16, 16 (figure), 64, 65 (figure)
    differences in opinions regarding, 65
    including in reports as findings, 64
    previous definition, 15 (figure)
    records review and, 35
historical research, 35–36, 36 (figure)
HMTUSA (Hazardous Materials Transportation Uniform Safety Act) of 1990, 3
HRECs. *See* historical recognized environmental conditions
HSWA (Hazardous Solid Waste Amendments) of 1984, 3

I
ICs (institutional controls), 16–17, 25, 26 (figure), 79, 91
indoor air quality, 71
initial site survey, 54–55
Innocent Landowner Defense (ILD), 2, 8, 19, 21, 89, 90 (figure)
institutional controls (ICs), 16–17, 25, 26 (figure), 79, 91
interior observations during site reconnaissance, 45–49
interviews
    content of, 60–61
    form for, 142–143
    with key site managers, 57
    with occupants, 60, 61 (figure)
    with owners or occupants neighboring abandoned properties, 57–58
    with past occupants, 57
    with past operators, 57
    with past owners, 57
    preferred sequence of events, 60 (figure)
    relationship and timing issues, 58–60
    requirements for, 29 (figure)
    with state and local government officials, 61–62
    user issues, 57
invitee safety, 55

J
joint and several liability, defined, 1

K
key site managers, interviews with, 57

L
Landowner Liability Protections (LLPs), 2, 8–9, 18 (figure), 79 (figure)
    low-level contaminants results, 121
    Phase I ESAs, 11, 19–20
    Phase II ESAs, 89–90, 91 (figure), 121
    responsibilities under AAI, 25
    users not seeking, 20

See also Bona Fide Prospective Purchaser defense; Contiguous Property Owner defense; Innocent Landowner Defense
laser-induced fluorescence (LIF), 110, 112 (figure)
lead, 71, 131
leaking underground storage tank (LUST) sites, 13, 16, 64–65, 67
legacy contamination, 25
legacy remedies, 25
lenders, defined, 8–9
liability, 1–2
LIF (laser-induced fluorescence), 110, 112 (figure)
likely release areas, 84, 84 (figure), 121
limitations
    of Phase I ESAs, 10, 11, 12 (figure)
    of Phase II ESAs, 80–81 (figure), 80–81, 93–94 (figure), 93–94
    of work plans, 102–103, 103 (figure)
limiting conditions during site reconnaissance, 37, 41 (figure)
LLPs. *See* Bona Fide Prospective Purchaser defense; Contiguous Property Owner defense; Innocent Landowner Defense; Landowner Liability Protections
local government officials, interviews with, 61–62
long format, Phase II ESA reports, 125, 127–129 (figure)
low-level contaminants results, 121
LUST (leaking underground storage tank) sites, 13, 16, 64–65, 67

M
magnetometer surveys, 110, 110 (figure)
mandatory physical setting records, 35, 35 (figure)
mandatory state and federal record information, 32–35
Manual of Analytical Methods (OSHA), 54
material threat, 20 (figure), 20–21, 64
meniscus, 112–113, 115 (figure)
migration, 35, 35 (figure), 71, 102, 114
minimum search distances, 32, 33 (figure), 64
mobile laboratories, 111, 115 (figure)
mold, 71, 85
multiwavelength delay fiber detection system, 111, 113 (figure)

N
National Contingency Plan (NCP), 2, 101
National Priorities List (NPL) sites, 89, 90–91 (figure)
no further action (NFA) letters, 65
no-detect results, 121
nonmandatory records. *See* optional records
nonscope considerations, Phase I ESAs, 71–73
    covered by other standards, 71–72
    examples of, 71
    including in reports, 64
    including recommendations in reports, 72
    special user needs, 72
nonscope considerations, Phase II ESAs, 131–132
    business judgments, 131–132
    contracting and subcontracting, 131, 132 (figure)
    issue-specific standards and regulations, 131
    legal risks, 132
    recommendations, 131
    safety and health, 131
NPL (National Priorities List) sites, 89, 90–91 (figure)
nuisance dumping, 40 (figure)

O
objectives
    including in Phase II ESA reports, 126, 126–127 (figure)
    of Phase I ESAs, 10, 10 (figure)
    of Phase II ESAs, 78–80, 79 (figure), 90 (table), 107
occupants, interviews with, 57–58, 60, 61 (figure)
Occupational Safety and Health Administration (OSHA), 53–54
operators, interviews with, 57
opinions, including in reports, 64–66
optional (nonmandatory) records
    physical setting records, 35, 35 (figure)
    state and federal records, 32–35
OSHA (Occupational Safety and Health Administration), 53–54

P
past owners, interviews with, 57, 142–143
personal protective equipment (PPE), 54
Phase I ESAs. *See* ASTM Standard E1527
Phase II assessors (consultants), 77–78, 84
    defined, 84 (figure)
    providing legal or business advice, 94
    reporting obligations and confidentiality, 95–96
    See also environmental professionals
Phase II ESAs. *See* ASTM Standard E1903
photography, 46 (figure), 48 (figure), 48–49
physical setting records, 35, 35 (figure)
Pollution Prevention Act (PPA) of 1990, 3
portable photoionization detectors (PIDs), 110, 110 (figure)
potentially responsible parties (PRPs), 1
PPE (personal protective equipment), 54
practically reviewable, 13–14, 31
producers, defined, 8–10
product liability, 65–66
Property Reconnaissance form, 135–138
PRPs (potentially responsible parties), 1
publicly available information, 31

Q
quality assurance/quality control (QA/QC), 83, 94
    data evaluation, 119–121, 120 (figure)
    report preparation, 129
    work plans, 102, 104

R

radon gas, 64, 71, 85
RCRA (Resource Conservation and Recovery Act) of 1976, 2, 11, 53, 101–102
reasonably ascertainable information, 20, 31, 32 (figure), 63
recognized environmental conditions (RECs)
- conclusions regarding, including in reports, 67, 67 (figure)
- de minimis concept, 16, 17 (figure)
- defined, 20 (figure)
- differences in opinions regarding, 64–65
- evaluation of data, 63
- identifying, 20–21
- including in reports as findings, 64
- recommendations regarding, including in reports, 67–68
- records review and, 35

recommendations
- including in Phase I ESA reports, 67 (figure), 67–68, 72
- including in Phase II ESA reports, 131
- verbal, 72
- written, 72

records
- historical, 35–36
- mandatory and optional physical setting records, 35
- mandatory and optional state and federal records, 32–35
- obtaining, 31–32
- reasonably ascertainable, 31
- sources of, 35 (figure)
- types of, 35 (figure)

records research radius maps, 34
RECs. *See* recognized environmental conditions
references, including in reports, 68
regulators, defined, 9–10
regulatory statutes, 2–3
- Asbestos Hazard Emergency Response Act (AHERA), 3
- Clean Air Act, 2
- Clean Water Act, 2
- Emergency Planning and Community Right-to-Know Act, 3
- Federal Water Pollution Act, 2
- Hazardous Materials Transportation Uniform Safety Act, 3
- Hazardous Solid Waste Amendments, 3
- Pollution Prevention Act, 3
- Resource Conservation and Recovery Act, 2, 11, 53, 101–102
- Secured Creditor Exemption, 3, 9, 22
- Superfund Amendments and Reauthorization Act, 3, 7, 22
- Toxic Substance Control Act, 2
- *See also* Brownfields Amendments; Comprehensive Environmental Response, Compensation, and Liability Act

relationship issues regarding interviews, 58–60
relevant experience, 28 (figure)

report preparation, Phase I ESAs, 63–69
- additional investigations, 66, 66 (figure)
- additional services, 67 (figure), 67–68
- appendices, 68–69
- conclusions section, 67, 67 (figure)
- contents of report section, 63
- data evaluation, 63
- data gaps, 66 (figure), 66–67
- deviations section, 68
- documentation section, 63
- environmental professionals' (EPs') statements, 68
- findings section, 64
- opinions on findings, 64–66
- references section, 68
- scope of services section, 63–64
- shelf life of reports, 23–24, 24 (figure)
- signatures, 68

report preparation, Phase II ESAs, 81, 109, 125–130
- activities section, 127 (figure), 129
- appendices, 129 (figure), 129–130
- background information, 127 (figure), 129
- conclusions section, 126 (figure), 128 (figure), 129
- contents of report section, 126–127 (figure)
- data evaluation, 126 (figure), 128 (figure)
- documentation section, 127–128 (figure), 129
- introductory components, 126, 127 (figure)
- long format, 125, 127–129 (figure)
- reporting obligations and confidentiality, 95–96, 96 (figure)
- required elements, 125, 126 (table)
- results section, 128 (figure), 129
- scope of services section, 126 (figure)
- short format, 125, 126 (figure)
- signatures, 126–127 (figure), 129
- specific language, 125
- written communication, 96

Resource Conservation and Recovery Act (RCRA) of 1976, 2, 11, 53, 101–102
responsibilities
- under AAI, 25–26
- of consultants, 97–99, 98 (figure)
- of environmental professionals (EPs), 28
- responsible charge, 28–29
- of users, 26–28, 63, 97, 98 (figure)

responsible charge
- attributes of, 28–29
- identifying environmental professional (EP) in reports, 63

results interpretation, Phase II ESAs, 121–123
- including in reports, 128 (figure), 129
- low-level contaminants results, 121
- no-detect results, 121
- unexpected results, 121, 122 (figure), 123

risk tolerance, 3, 9, 29, 67, 80, 93

S

safety and health
- health and safety plans, 104
- initial site survey, 54–55
- of invitees, 55
- as nonscope issue in Phase II ESAs, 131
- overview, 51
- site safety plans, 54
- standard operating procedures, 53–54
- of subcontractors, 55
- training requirements, 51–53

sample collection, 80, 94, 120–121
- environmental media sampling, 111–112
- representativeness, 103 (figure)
- sample handling, 112–116

sampling plans, 103–104 (figure), 103–104, 108–109
SARA (Superfund Amendments and Reauthorization Act) of 1986, 3, 7, 22
scientific method, 77, 83, 83 (figure), 89, 90 (figure)
scope, of Phase I ESAs
- limitations of standard, 10, 11
- objectives of standard, 10
- purpose of standard, 10
- statement of, 10–11, 11 (figure)

scope, of Phase II ESAs, 77–81, 78 (figure)
- limitations of standard, 80–81
- objectives of standard, 78–80
- users' needs, 80, 92 (figure)

scope, of work plans, 101–102
scope of services
- including in Phase I ESA reports, 63–64
- including in Phase II ESA reports, 126 (figure)

Secured Creditor Exemption, 3, 9, 22
shelf life of reports, 23–24, 24 (figure)
shipping box design for soil sample transport, 114, 116 (figure)
short format, Phase II ESA reports, 125, 126 (figure)
signatures
- including in Phase I ESA reports, 68
- including in Phase II ESA reports, 126–127 (figure), 129

site reconnaissance, 37–38
- exterior observations, 43–45
- general observations, 42–43
- general site setting, 38–40
- interior observations, 45–49
- observable conditions, 38 (table)
- requirements for, 29 (figure)

site safety plans, 54, 54 (figure)
site visits, defined, 37
Small Business Liability Relief and Brownfields Revitalization Act of 2002. See Brownfields Amendments
standard operating procedures (SOPs), 53 (figure), 53–54
state government officials, interviews with, 61–62

strict liability, defined, 1
subcontractors, 97, 97 (figure)
- as nonscope issue in Phase II ESAs, 131, 132 (figure)
- safety and health, 55

Superfund Amendments and Reauthorization Act (SARA) of 1986, 3, 7, 22
Superfund Law. See Comprehensive Environmental Response, Compensation, and Liability Act

T

target analytes
- conceptual model, 108, 108–109 (table)
- detection problems, 93
- results interpretation, 121
- threshold levels, 94
- validation process, 119, 120 (figure)

technical photography, 49
terminology, Phase I ESAs, 13–18
- All Appropriate Inquiries Final Rule, 15, 15 (figure)
- reasons for, 13–15
- sources of related, 17–18

terminology, Phase II ESAs, 83–85, 86–87 (figure), 88 (table)
timing issues regarding interviews, 58–60
Toxic Substance Control Act (TSCA) of 1976, 2
training requirements, 51–53, 53 (figure)
Transaction Screen (TS). See ASTM Standard E1528

U

underground storage tank (UST) systems, 35, 85, 92 (figure)
- evidence of repair, 45 (figure)
- leaking, 13, 16, 64–65, 67
- vent lines, 58, 59 (figure)

unexpected results, 121, 122 (figure), 123
User Questionnaire form, 27 (figure), 63, 139–140
User Responsibility Statement form, 141
users
- interviews with, 57
- needs of, 80, 80 (figure), 92 (figure)
- not seeking LLPs, 20
- overview, 8–10
- responsibilities of, 26–28, 141
- special needs of, 72
- user questionnaire, 27 (figure), 63, 139–140

UST. See underground storage tank systems

V

validation process, Phase II ESAs, 119–121, 120 (figure)
vapor migration and intrusion, 35 (figure), 71
voluntary use of E1527 standard, 19

W

waste generation, 97, 97–98 (figure)
windfall liens, 25–26

work plans, 101–105
    chemical testing plan, 104
    data quality objectives, 102
    defined, 101
    health and safety plan, 104
    limitations of, 102–103, 103 (figure)
    potential distributions of contaminants, 103
    quality assurance/quality control plan, 104
    review of existing information, 102
    sampling plan, 103–104
    scope of, 101–102
    statement of objectives, 101

X

X-ray fluorescence (XRF), 110, 111 (figure)